Communications
in Computer and Information Science
2085

Editorial Board Members

Rationale

The CCIS series is devoted to the publication of proceedings of computer science conferences. Its aim is to efficiently disseminate original research results in informatics in printed and electronic form. While the focus is on publication of peer-reviewed full papers presenting mature work, inclusion of reviewed short papers reporting on work in progress is welcome, too. Besides globally relevant meetings with internationally representative program committees guaranteeing a strict peer-reviewing and paper selection process, conferences run by societies or of high regional or national relevance are also considered for publication.

Topics

The topical scope of CCIS spans the entire spectrum of informatics ranging from foundational topics in the theory of computing to information and communications science and technology and a broad variety of interdisciplinary application fields.

Information for Volume Editors and Authors

Publication in CCIS is free of charge. No royalties are paid, however, we offer registered conference participants temporary free access to the online version of the conference proceedings on SpringerLink (http://link.springer.com) by means of an http referrer from the conference website and/or a number of complimentary printed copies, as specified in the official acceptance email of the event.

CCIS proceedings can be published in time for distribution at conferences or as post-proceedings, and delivered in the form of printed books and/or electronically as USBs and/or e-content licenses for accessing proceedings at SpringerLink. Furthermore, CCIS proceedings are included in the CCIS electronic book series hosted in the SpringerLink digital library at http://link.springer.com/bookseries/7899. Conferences publishing in CCIS are allowed to use Online Conference Service (OCS) for managing the whole proceedings lifecycle (from submission and reviewing to preparing for publication) free of charge.

Publication process

The language of publication is exclusively English. Authors publishing in CCIS have to sign the Springer CCIS copyright transfer form, however, they are free to use their material published in CCIS for substantially changed, more elaborate subsequent publications elsewhere. For the preparation of the camera-ready papers/files, authors have to strictly adhere to the Springer CCIS Authors' Instructions and are strongly encouraged to use the CCIS LaTeX style files or templates.

Abstracting/Indexing

CCIS is abstracted/indexed in DBLP, Google Scholar, EI-Compendex, Mathematical Reviews, SCImago, Scopus. CCIS volumes are also submitted for the inclusion in ISI Proceedings.

How to start

To start the evaluation of your proposal for inclusion in the CCIS series, please send an e-mail to ccis@springer.com.

Paulin Melatagia Yonta · Kamel Barkaoui ·
René Ndoundam · Omer-Blaise Yenke

Editors

Research in Computer Science

6th Conference, CRI 2023
Yaounde, Cameroon, December 12–13, 2023
Proceedings

Springer

Editors
Paulin Melatagia Yonta
University of Yaoundé I
Yaoundé, Cameroon

Kamel Barkaoui
Conservatoire National des Arts et Métiers
Paris, France

René Ndoundam ⓘ
University of Yaoundé I
Yaoundé, Cameroon

Omer-Blaise Yenke ⓘ
University of Ngaoundéré
Ngaoundéré, Cameroon

ISSN 1865-0929   ISSN 1865-0937 (electronic)
Communications in Computer and Information Science
ISBN 978-3-031-63109-2   ISBN 978-3-031-63110-8 (eBook)
https://doi.org/10.1007/978-3-031-63110-8

This Springer imprint is published by the registered company Springer Nature Switzerland AG
The registered company address is: Gewerbestrasse 11, 6330 Cham, Switzerland

If disposing of this product, please recycle the paper.

Preface

This volume of Communications in Computer and Information Science (CCIS) contains the papers selected for presentation at the sixth edition of the Conference on Research in Computer Science –Conférence de Recherche en Informatique – CRI 2023, held from 12 to 13 December 2023 at the Department of Computer Science, Faculty of Sciences of the University of Yaounde I, Yaounde, Cameroon.

Since its foundation in 2013 by Maurice Tchuente, CRI is Central Africa's leading conference where young and experienced researchers in computer science meet to discuss issues, challenges, opportunities, and risks and present innovations and scientific results. It has attracted innovative works and followed closely the developments of the trending fields of computer science.

The current conference proceedings reflect the principal fields for the computer scientists of Central Africa and particularly of Cameroon: Artificial Intelligence, Machine Learning, Natural Language Processing, Computer Vision, Cryptography and Distributed Computing.

The sixth edition of CRI had submissions from 9 countries (Algeria, Burkina Faso, Cameroon, Central African Republic, France, Germany, Luxembourg, South Africa and the USA) and from 21 universities. The accepted papers were selected from 72 submissions, each of which was reviewed by at least three reviewers using the double-blind mode. The Program Committee co-chairs and the scientific secretary had two consultation meetings to look at all the reviews and made the final decisions; 16 papers were finally accepted.

For this edition of the conference, we had the opportunity to welcome three invited presentations:

- Gérald Ekembe Ngondi (Lero, the SFI Research Centre for Software, Ireland), Formal Methods and Programming Language Translation – Translating CCS into CSP
- Omer Nguena Timo (Université du Québec en Outaouais, Canada), Modéliser et tester des systèmes évolutifs
- Didier Donsez (Université Grenoble Alpes GINP, France), L'Internet des Objets (Isolés) par Satellite (SatIoT)

We would like to thank all authors who submitted papers for review and for publication in the proceedings and all the reviewers for their time and effort to provide the reviews on time.

We are also grateful to the Faculty of Sciences of the University of Yaounde I, UMMISCO/IRD, ASDS, FR2I, INRIA, CAYSTI and CAIS for sponsoring the conference.

April 2024

Paulin Melatagia Yonta
Kamel Barkaoui
Rene Ndoundam
Omer-Blaise Yenke

Organization

Scientific Secretary

Paulin Melatagia Yonta — University of Yaoundé I, Cameroon

Program Committee Chairs

Kamel Barkaoui — Conservatoire National des Arts et Métiers, France
Rene Ndoundam — University of Yaoundé I, Cameroon
Blaise-Omer Yenke — Université de Ngaoundéré, Cameroon

Program Committee

Alain Tchana — Grenoble INP - Ensimag, France
Armel Fotsoh — Yxir-EDF Group, France
Arouna Ndam Njoya — University Institute of Technology of Ngaoundéré, Cameroon
Blaise-Omer Yenke — IUT, University of Ngaoundéré, Cameroon
Belgacem Ben Hedia — Commissariat à l'Energie Atomique-LIST, France
David Jaures Fotsa Mbogne — Université de Ngaoundéré, Cameroon
Djamil Aissani — University of Bejaia, Algeria
Franklin Tchakounte — University of Ngaoundéré, Cameroon
Fritz Mbounja Besseme — University of Ngaoundéré, Cameroon
Gayo Diallo — University of Bordeaux, France
Georges Edouard Kouamou — École Nationale Supérieure Polytechnique de Yaoundé, Cameroon
Gerard Ekembe Ngondi — Lero SFI Research Centre for Software, Ireland
Ghislain Auguste Atemezing — Mondeca, France
Hayatou Oumarou — Université de Maroua, Cameroon
Hyppolite Tapamo — Université de Yaoundé I, Cameroon
Innocent Souopgui — University of New Orleans, USA
Jean Etienne Mboula Ndamlabin — Inria Nancy - Grand Est, France
Jerry Lonlac — IMT Nord Europe, France
Justin Moskolai Ngossaha — University of Douala, Cameroon

Kamel Barkaoui	Conservatoire National des Arts et Métiers, France
Kengne Tchendji Vianney	University of Dschang, Cameroon
Louis Fendji	Université de Ngaoundéré, Cameroon
Louis Fippo Fitime	École Nationale Supérieure Polytechnique de Yaoundé, Cameroon
Malo Sadouanouan	Université Polytechnique de Bobo-Dioulasso, Burkina-Faso
Marcellin Nkenlifack	Université de Dschang, Cameroon
Martin Luther Mfenjou	University of Ngaoundéré, Cameroon
Maurice Tchuente	Université de Yaoundé I, Cameroon
Max Pambe	University of Maroua, Cameroon
Mohamed Ghazel	Université Gustave Eiffel, France
Norbert Tsopze	Université de Yaoundé I, Cameroon
Patrice Bonhomme	Université François Rabelais de Tours, France
Paul Dayang	University of Ngaoundéré, Cameroon
Paulin Melatagia Yonta	University of Yaoundé I, Cameroon
Rabah Ammour	Aix-Marseille Université, France
Remy Maxime Mbala	University of Ngaoundéré, Cameroon
Rene Ndoundam	University of Yaoundé I, Cameroon
Rodrigue Domga Komguem	University of Yaoundé I, Cameroon
Samir Ouchani	CESI Lineact, France
Samuel Bowong	University of Douala, Cameroon
Sylvain Iloga	University of Maroua, Cameroon
Vivian Nwaocha	National Open University of Nigeria, Nigeria
Vivient Corneille Kamla	University of Ngaoundéré, Cameroon

The following colleagues also reviewed papers for the conference and are due our special thanks: Hervé Tale Kalachi and Marsien Ayemedie Moto

Organization Committee

Aminou Halidou	University of Yaoundé I, Cameroon
Etienne Kouokam	Université de Yaoundé I, Cameroon
Hyppolite Tapamo	Université de Yaoundé I, Cameroon
Norbert Tsopze	Université de Yaoundé I, Cameroon
Paulin Melatagia Yonta	University of Yaoundé I, Cameroon
Rene Ndoundam	University of Yaoundé I, Cameroon
Samuel Bowong	University of Douala, Cameroon
Blaise-Omer Yenke	IUT, University of Ngaoundéré, Cameroon

Contents

Convolutional Neural Network Based Detection Approach of Undesirable SMS (Short Message Service) in the Cameroonian Context

Loic Youmbi[1,3]([✉]) [iD], Ali Wacka[2] [iD], and Norbert Tsopze[1,3] [iD]

[1] Department of Computer Science, Faculty of Sciences, University of Yaounde 1, Yaoundé, Cameroon
loicyoumbi9@gmail.com
[2] Department of Computer Science, Faculty of Sciences, University of Buea, Buea, Cameroon
ajoanberi@gmail.com
[3] Sorbonne Universite, IRD, UMMISCO, 93143 Bondy, France
tsopze.norbert@gmail.com

Abstract. With the low cost of mobile phones, SMS(Short Message Service) and the advent of communications software (whatsapp, telegram, etc.), many people communicate easily. Many commercial companies, entities or individuals also use SMS to send out advertisements, unwanted messages containing links or malicious content that may violate customer privacy. This raises the question of how to help users to avoid being trapped? Several works have been proposed to counter these unwanted messages (SMS SPAM). Most of these are only tested on messages written in English, are not adaptable to messages from bilingual countries and are based on datasets dating back to 2012. Since Cameroon is a bilingual country, in order to set up a model taking into account SMS written in French, English or mixed language, we create a Cameroonian SMS spam dataset. The aim of this work is to propose a convolutional neural networks(CNN) based model for the processing of the Spam SMS dataset in the Cameroonian context to classify a given SMS as SPAM or HAM. The results obtained after experimentation in terms of accuracy, precision, recall and F1-score gave respectively 99.59%, 98.3%, 98%, 98.1%.

Keywords: Convolutional Neural Network · SMS · SPAM SMS · HAM SMS · Detection

1 Introduction

The advancement of ICT (Information and Communication Technology) has allowed many people to own a smartphone. We observe an increased rise in the use of ICT tools by young people. In the smartphones age, user has confidential and personal information such as passwords, images, numbers of credit card, contact lists that stored on these phones, making those users more vulnerable

P. Melatagia Yonta et al. (Eds.): CRI 2023, CCIS 2085, pp. 1–14, 2024.
https://doi.org/10.1007/978-3-031-63110-8_1

to cyber attacks by spam SMS (Shorts Messages Services) [15,17] ... It is in this sense that many companies or entities prefer to use this service to better communicate about their services or products with customers. [4] claims that the average volume of SMS sent per month increased by 7700% between 2008 and 2018.

SMS (Short Message Service) is a text message, sent by a user, a company or a phone operator to one or more people either to communicate, to pass advertisements, to pass an information, ... This service is characterized by: a sender, the receiver(s), the sending date, the receiving date, and a content (text). The use of SMS compared to emails does not require an internet connection, but only a network coverage. Therefore, the user can directly start communicating through it. When an SMS arrives, the receiver is directly notified and can read it without any incident, whereas compared to emails, the user has to connect to the internet in order to receive the email. According to [13], SMS have become a multi-billion dollar business. Due to the low cost of SMS in most telecommunication service providers, as well as its accessibility and efficiency compared to email services, it is one of the most common communication tools in the world [5].

Spam is an undesired text message sent to the user's phone with different types of content. Some objective of spammers is to steal critical information about the user with different types of content, such as advertisements, rewards, free services and promotions [1]. This information can be: username, password and credit card data. Spam contains unwanted information, malicious URL links, URLs that link to malicious sites to download malware or steal personal information for identity theft. Detection of SMS spam seems to be difficult due to a limited set of characters, lack of cost effective SMS spam datasets, different dialects. The problem arises is therefore to implement an SMS spam detection tool for helping users to avoid the traps of spammers. Several works have been proposed to counter unwanted messages (SMS SPAM) [12] [11] [10]. Except that the methods used have shortcomings especially when we go back to the Cameroonian context with the two officials languages, dialects and also services offered by SMS. Most of the existing methods are based on SMS written only in English. The paper presents a framework of spam detection in Cameroonian context. The main steps to achieve this objective are: (1) we create a multilingual SPAM SMS dataset by collecting users SMS in from our phones and some groups in social media, (2) the collected SMS are preprossessed by removing stop words and replacing some terms with meaningful words, (3) The resulting text is embedded in numerical representation, (4) a designed CNN (Convolutional Neural Network) model is trained to classify as spam or not. Training the model using the Cameroonian dataset permits to avoid the geographical bias due to the out of context collected data.

The Sect. 2 will deal with the related work on SMS spam detection. In Sect. 3, the methodology of machine learning algorithm for spam detection is presented and in Sect. 4 the experiment results.

2 Related Work

This section focuses on some research works about spam detection. The works include the machine learning and deep learning spam detection based approaches.

In [3], an undesirable SMS spam detection system is proposed to identify an efficient set of features using Restricted Boltzmann Machine (RBM). The proposed framework classifies SMS spam and ham using a Deep Neural Network (DNN) classifier. They use data from the UCI Machine Learning Warehouse which was created in 2012 [2]. Preprocessing step has to be performed using unsupervised filters like replace the missing values, remove duplicates, etc., and the min-max normalization system is used to normalize data between [0,1]. The experimentation comes about on SMS spam dataset demonstrates that DNN can learn a better generative model and perform well on SMS spam recognition task.

In [6], authors present a dynamic deep model for spam detection that adjusts its complexity and extracts features automatically. They use convolutional and pooling layers for feature extraction as well as basic classifiers such as random forests and extremely random trees to classify texts as spam or legitimate. The data goes through two main phases: they apply the word embedding technique after prepossessing to convert the text data into a digital form and use deep convolutional forest (DCF) to extract features and classify the text. The proposed method analyses the SMS Spam Collection Data set from UCI [16] and outperformed existing traditional machine learning classifiers and deep neural networks and achieved the highest accuracy rate of 98.38%.

In [7], authors propose various models with mixed text in Arabic's or English's written messages that are collected manually based on traditional machine and on deep learning algorithms. The proposed system starts by cleaning up unnecessary information found in the text messages. Subsequently, a prepossessing task will be applied to represent the text data in a compatible form that will serve as input to the machine or deep learning methods. After completing the data preparation, classification algorithms is applied in order to distinguish spam from non-spam messages. The experimental evaluation of the proposed approach showed that the CNN-LSTM model outperforms other SMS spam classification techniques.

In this research [14], authors propose a new SMS spam detection technique based on the study of English language SMS spam cases using natural language process and deep learning techniques. In order to prepare the data for the model development process, the authors use word tokenization, data padding, data truncation, and word completeness to give more dimension to the data from [16]. This data is used to develop models based on LSTM and GRU algorithms. The performance of these models is compared to models based on (SVM, NB). This model achieves an overall accuracy of 98.18% and detects SPAM messages with an accuracy of 90.96% and an error of 0.74%.

In [8], authors have implemented a special architecture known as Long Short Term Memory (LSTM) using the SMS Spam Collection dataset [16] Before using LSTM for classification, data is converted into semantic word vectors using word2vec, WordNet and Concept Net. The classification results are compared with the benchmark classifiers like Support Vector Machine, Naive Bayes, ANN, kNN and Random Forest. The results obtained are evaluated using metrics like accuracy and F-measure and improves state of the art methods results.

From the above, we can see that the models mentioned in the state-of-the-art have the following limitations:

- Most of the state-of-the-art's datasets are old (2012) and do not fit the current context anymore. Nowadays SMS systems have expanded into emojis, the mobile money concept with different services.
- many of these datasets are written only in English or only in one language and may have limitations on data with mixed languages.
- These data are not collected in Cameroon. This could introduce bias in the model due its out of context aspect.
- other models remove punctuation, mathematical symbols, special characters, ... without prepossessing words such as: URL, phone number, currency symbols, e-mail address, ... we believe that the removal of these terms makes the information about the SMS in question lost.

3   Methodology

In order to take into account some of the limitations mentioned in the state of art we have set up a dataset named **cameroun_spam** available on the link below[1]. To show how representative the dataset is, we experiment with a deep learning algorithm. This allows us to build a deep learning model for the detection of unwanted SMS in the Cameroonian context.

3.1   Data Collection and Description

Messages from telephone operators come from several sources. First of all, we created a Google form and asked anyone of goodwill to send messages to it. The messages received were more SMS spam, which telephone operators use to pass on advertisements, and messages from Cameroonian scammers. However, another part was made up of messages from the government to raise awareness about road traffic, hate messages, national unity, education, sexuality, etc. In the end, we had around 8556 SMS messages. Given the small size of the latter, in order to take into account several contexts, we collected messages from WhatsApp groups. Simple SMS (HAM) were collected through SMS conversations and also in some social media groups. Thus, a message is considered HAM if it is sent by a telephone operator and also if it comes from the conversations of known users, whatsapp groups, or also the government without any danger. In

---

[1] https://docs.google.com/spreadsheets/d/153G4rxniGr-6Lbu1g8kyPFZI99hYUeaM/edit?usp=sharing&ouid=101356100199235048237&rtpof=true&sd=true.

order to get closer to the victim of SMS spam attacks, we collected data from most of the phones of people aged 15 to 45 and also from the wathsapp groups of english students, french students, job seekers, immigrants, association groups, teachers groups, social development groups. These SMS include different subjects such as studies, love, finance, telecom service, health, ... For the simples SMS, we have largely collected the SMS from the most used telephone operators in Cameroon. Also, most of the spam messages are based on SMS from electronic money services (Mobile and Orange Money). A SMS is considered SPAM if it is sent by a third party to pass an advertisement, redirects to an unknown site offering earnings or promotions and those of the telephone operators who run advertisements. Basically, we have a dataset labelled Ham or Spam consisting of 23739 SMS of which 2880 are Spam SMS and 20859 are Ham.

3.2 Proposed Methods

Fig. 2 presents the workflow of the designed methodology. This workflow will be described in this section.

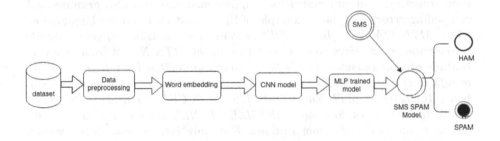

Fig. 1. General Flow of Main Phases for model Task.

3.3 Data Analysis

We also analyze the number of times a word occurs in a sentence. Generally speaking, most words have a minimum of 2 and a maximum of 6 occurrences. In general, there are prepositions (stop word). For some particular terms, we have words like MTN, Orange, Bonus, camtel, Nexttel, FCFA,...

Data Structure. Given that the primary objective is to set up an SMS Spam detection model taking into account the Cameroonian context, the dataset set up includes several languages such as: French, English, French and mixed English and other terms relating to the Cameroonian context. Cameroon being a bilingual country, many people communicate in these two languages and are sometimes tempted to mix them in the same conversation.

- **English Language:** HAM messages from telephone operators are written in everyday language and do not contain colloquial terms. Ham messages from ordinary users are written in colloquial and/or everyday language and may contain contextualised terms. For example :*i love you* can still be written *i love u* or *i ve u* or even *Ask your mother or your father* can still be *Ask ur mother or ur father*. As a result, these strange terms are preserved and should be studied in future work (explicability of neural networks).

For SPAM messages, those from telephone operators are in a familiar language because, for them, the aim is to reach the maximum number of customers depending on the time of year, the current context and the new fashionable jargon. For example: *Download a waiting ZIK at 150u/month by sending 5 or 6 or 7 to 8706. 5. It is not normal to Toofan 6. Child of God by Scanner Neville 7. Mokodo by Serge Beynaud See more at *194#*. The following is an example of SPAM operators using colloquial language: *Yo! na how? This Wednesday, the gigas are back in full swing. Subscribe to Pulse FapCent+ via My Orange or #119*20#: 500U = 3.4GB valid until 23H59.*

Unlike operator spam, which is mostly written in colloquial language, SPAM messages from scammers are mostly written in strong language, although they sometimes contain French or English mistakes and also grammatical or spelling errors. A typical example of Spam written in strong language is: *8787<MTN WANDA> BRAVOO!!!dear customer, considering your multiple transactions made since your subscription to the MTN NETWORK, the percentage of your consumptions is 80%. Your number has been drawn and you benefit from a bonus MTN of a value of [260. Your number has been drawn and you will receive an MTN bonus worth [260. 000FCFA]+ 1 Android 4G smart phone to be received from your MOBILE MONEY account and to be withdrawn in all total and bocom stations. For more information please contact Mrs MATENE beatrice on the standard number [679 20 78 38] for the transfer of your bonus. MTN thanks you for your loyalty. www.http//winning.COM.* We can see that spam messages from spammers are very long and for the most part contain mathematical terms such as: %,+, .. and also exclamations. Their main objective is to pass themselves off as a telephone operator, which is why they use the term *8787* followed by *MTN WANDA*.

Spam messages from wathsApp groups are mostly messages containing redirection links to unknown sites. These contain terms to convince users to visit the unknown sites and also to share the message in several groups or with several users. They are written in everyday language and depend on the time of year or country context. An example of this type of spam is:

PAUL BIYA OFFERS 40GB OF FREE DATA AND 1000CFA TO HIS SUPPORTERS TO CELEBRATE HIS BIRTHDAY*. The Biya organisation is joining forces with all network providers. Offering 40GB of free data and a 1,000 CFA airtime voucher to his supporters.
I've just received mine, take yours below
https://br.ke/Paul-Biya-cadeau-d *birthday*
In summary, we can see that the most frequently used Cameroonian contextu-

alised English terms in SMS messages are:*u, yo, offers, free, win, na, ur, benefits, data, airtime, receive, take, birthday, clicking, free data, fcfa, phone number, Bonus, njooh, bonus code, name, payment, Back2schoolBonus, ...*

- **French language** As far as messages written in French are concerned, the language used by operators is also strong when it comes to HAM messages and rather colloquial when it comes to SPAM messages. For simple users, they are written in colloquial language and more contextualised according to the type of communication (discussion) and the interlocutor. The following is an example of an operator spam message: *Super surprise 651498065! Tapez *204*6# Ce bonus est pour vous! Profitez de MTN Udaily et amusez-vous sans limites! Cout: 100F/jour.*

The following is an example of a ham message from a simple user:*Bsr svp je quelqu'un aurais le numéro du brun de la classe j'ai pas son nom . Svp*

Just as messages from spammers written in English are in supported language, spam messages from spammers written in French are also in the same language. The aim remains the same: to persuade users to share the message, to convince them to click on a link or contact a number to take advantage of a fictitious offer. In fact, these messages contain more words than the hams messages and contain terms such as: *<MTN WANDA> BRAVOO!!chèr(es) client(es), vus vos multiples transactions éffectuées dépuis votre abonnement AU RESEAU MTN, le pourcentage de vos consommations est de 86%. Votre numéro à été tiré au sort et vous bénéficiez d'une prime bonus MTN d'une valeure de 245.000FCFA + 1 WELCOME PACK Android 4G à recevoir a partir de votre compte MOBILE MONEY et a rétirer dans toutes les stations total et bocom.pour plus d, informations veuillez contacter Mr Pokam Franck au numéro standard 670 18 16 63 pour le virement de votre prime bonus. MTN vous remercie pour votre fidélité.*www.http//winning.COM or even *Le tchio de décembre c'est ce 22/12 au stade Omnisport de Bépanda. Kizz Daniel sera là. Tickets disponibles ici :*

Android: https://bit.ly/AppOrangeMoney_Android

SPAM messages received via social networks are also written in everyday language and have the same characteristics as those written in English. The objectives remain the same as those written in English, i.e. to encourage the user to follow a malicious link, to share messages in several groups or with several users). Below is an example of a spam SMS written in French and sent to whatsapp groups: *Urgent! pour tous les camerounaise* MTN - Orange - Camtel sont heureux d'offrir à tous 6000 francs de crédit d'appel 25Go internet gratuit disponible pendant un mois *Orange* 6uil.com/cameroon?Orange *Camtel* 6uil.com/cameroon?Camtel*

In conclusion, the terms most commonly used in messages written in French depending on the Cameroonian context are: *Urgent, camerounais, offrir, credit, internet, pourcentage, multiple, prime, station, BRAVOO, chers, client, mtn, orange, camtel, nexttel, nestle, whatsapp, Carimo, RDPC, cc, bby, bjr, Cc, pcq, ufkw, oki, go, gars, truck, ...*

Occurrence and Replacement Characters

- *MonnaieSymbol ou SymboleMonnaie.* For each SMS, we set up a regular expression to detect terms beginning or ending with FCFA, fcfa, FR,fr, F,f, XFA, Xfa, U, u, million,milliard. When we encounter these, we replace them with moneySymbol if the message is written in English, or SymboleMonnaie if it's written in French. Statistically, over 2408 SMS contain these terms.
- *PhoneNumber ou NumeroPhone.* In the same way as in the previous terms, we replace the telephone numbers found in messages with the terms phoneNumber or numeroPhone, depending on whether the message is written in French or English.

  Statistically, these terms appear more than 611 times in the data set. The most common terms are: 00237679196389 / 00237 6 79 19 63 89/ 00237 679 196 389/ +237679196389/ +2376 79 19 63 89/ +237 679 196 389/ @237679196389/ @237 6 79 19 63 89/ @237 679 196 389/ 679196389/ 679 196 389/ 6 79 19 63 89/ 6-79-19-63-89/6 79 19 63 89/679 196 638/@679196389/@6 79 19 63 89/ @6-79-19-63-89...
- *Data connection.* As mentioned above, we scan the data, and for each message, we check the presence of terms such as GB, Gb, MB, Mb, MO, Mo, GO, Go, mo, mb,... we have listed 1481 occurrences. We replace these terms with data Example: 100GB→ data, 10Go→ data
- *Web address or HTTP.* In order to avoid the loss of web address information when deleting punctuation marks in a sentence, for example, we've sentence, we've created a regular expression that extracts in a given SMS. To be more precise words beginning with http://n_importe_quoi or https://n_importe_quoi are indexed and replaced by the following terms webaddress or webAddress, depending on the language of the SMS. In the end, we have 1946 SMS messages in which these terms appear, and according to the number of according to the number of occurrences, we noticed that they appear only once in a given SMS. The diagram below summarizes the different occurrences.
- *The codes.* Still with a view to keeping certain terms as they are without loss of information, we also decided to set up a regular expression to extract the codes found in SMS messages. The codes found in our dataset are : *123#, * 123 * 11#, *123*111#, * 123 * 111*22#, *123 * 1 * 2 #, ... we noticed that most of the codes found are for the operators MTN/ORANGE and also Camtel. Out of 23,700 SMS, we found 134 SMS bearing the codes in question, and based on the number of occurrences per SMS, we drew up the correspondence table below:

Representation Table. All in all, we counted 6735 SMS containing at least one of the following special symbols: web or HTTP address, monneySymbol, data, code, phoneNumber, ... After detecting and replacing the key terms, a spam message with the initial content: *8787<MTN WANDA> BRAVOO!!!dear customer, considering your multiple transactions made since your subscription to the MTN*

Table 1. Summarise table of data analysis.

occurrences by SMS	1 time	2 times	3 or more time
Million/milliard	45	5	0
u/U	676	123	42
FCFA/fcfa/fr/xfa/cfa...	915	139	167
Phone Number	514	94	3
Data Connection	1341	93	47
Web address	1946	–	–
Code data	487	20	1

NETWORK, the percentage of your consumptions is 80%. Your number has been drawn and you benefit from a bonus MTN of a value of [260. Your number has been drawn and you will receive an MTN bonus worth [260. 000FCFA]+ 1 Android 4G smart phone to be received from your MOBILE MONEY account and to be withdrawn in all total and bocom stations. For more information please contact Mrs MATENE beatrice on the standard number [679 20 78 38] for the transfer of your bonus. MTN thanks you for your loyalty. www.http//winning.COM will now have the following content: *8787<MTN WANDA> BRAVOO!!!dear customer, considering your multiple transactions made since your subscription to the MTN NETWORK, the percentage of your consumptions is 80%. Your number has been drawn and you benefit from a bonus MTN of a value of [260. Your number has been drawn and you will receive an MTN bonus worth [MonnaieSymbol]+ 1 Android 4G smart phone to be received from your MOBILE MONEY account and to be withdrawn in all total and bocom stations. For more information please contact Mrs MATENE beatrice on the standard number [PhoneNumber] for the transfer of your bonus. MTN thanks you for your loyalty. Webaddress* Having extracted all these terms from an SMS, here's an example to illustrate:

Representation Diagram

3.4 Data Prepossessing

The first step consists in: for each given message, the lemmatization technique is used to transform each word to its initial term. For example: eaten, eating become eat. Then, regular expressions are applied to detect and transform email addresses, web addresses, phone numbers and numbers respectively into *emailaddress, webaddress, phonenumber* and *number*. This is to ensure that the key terms in the spam messages do not disappear when we remove stop words and punctuation. After this phase, punctuation are eliminated. When a word contains a punctuation mark, we remove the punctuation and delete the space created. Since the dataset contains SMS written either in French or in English,

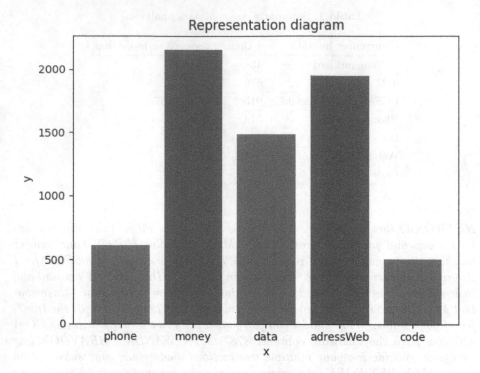

Fig. 2. Representation diagram of data analysis.

we used a stop word deletion technique. This technique is as follows: first, we use an already trained model (one of the fasttest library), for each message, we first detect its language and then use the stop word of nltk.corpus to delete it according to whether the language of the text is French, English or mixed. This technique of removing stop words not only allows the model to focus on the most relevant terms in each SMS, but also reduces the size of each message: initially the maximum size of message is 227 words, after full prepossessing the maximum size is 144 words.

3.5   Words Embedding

This part consists in fine-tuning a word2vec model on the dataset in order to use it in the following to convert the textual data into numerical data that can be passed to the input of a neural network. Since the model uses a neural network, the data obtained after the prepossessing phase (for each given message) will be converted into numerical data. For this purpose, in order to take into account the semantic dependence of the words of a message, a skip-gram model was fine-tuned by taking the groups of 5 g.

3.6 Convolutional Neural Network

The convolutional neural network used here is described as follows:

- **The convolutional layer** It works like a feature extractor by performing template matching by applying convolution filtering operations. The first layer filters the message with several convolution kernels, and returns feature maps, which are then normalised and/or resized.
- **Pooling layer** It is often placed between two convolution layers and receives several feature maps as input, applies the pooling operation to each of them. This consists of reducing the size of the messages, while preserving their characteristics.
- **The ReLU correction layer** This layer is made up of a set of ReLU units and is used to operate mathematical functions on the output signals. ReLU (Rectified Linear Units) refers to the real non-linear function defined by ReLU(x)=max(0,x).
- **Flattern layer:** The final stage in the "information extraction" section, flattening, simply consists of putting all the features (matrices) obtained end-to-end to form a (long) vector.

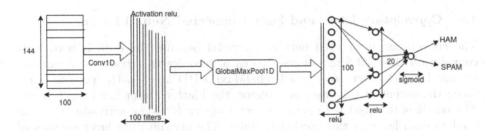

Fig. 3. General Flow of Main Phases for CNN Task.

3.7 Fully Connected Neural Layer

The fully-connected layer is used to classify a message at the input of the network by assigning a message to a class involves multiplying the input vector by the matrix containing the weights. It takes as input the output of the flatten layer and classifies as spam or ham.

4 Experimental Protocol

We have trained the model on google colab, 64bit processor, 12.68GB Ram. The execution takes about 20 min.

Table 2. Summarise table of result in our dataset and UCI dataset.

Dataset	SPAM	HAM	ration of Spam	ratio of Ham	Total
our dataset	3068	17562	14.87%	85.16%	20630
UCI dataset	747	4825	13.40%	86.59%	5572

4.1 Data-Set Ratio

4.2 Word Splitting, Word Embedding

The first step is to separate the data from the labels because it is a supervised learning model. For this experimentation, 80% of the data is used in the training phase and 20% in the test phase. Since the SMS messages are of different sizes, and the model takes as input data of uniform size, we set the size of each message to 144 words. For those messages whose size is less than this number, we make a padding operation consisting in filling it with zeros. The size of each matrix is thus 100*144, where 100 is the size of the representation vector of a word in a message and 144 is the number of words in the message.

4.3 Convolution Layer and Fully Connected Neural Layer

The convolution layer used here is a conv1D because the data is text. This convolution uses 100 filters of size 100 each. In order to extract the maximum feature for each filter, we use a GlobalMaxPool1D layer in the pooling layer. Since the output of this layer is a vector, the Flattern layer has been ignored. The result of this part is therefore a vector of size 100. The activation function used in each layer in this section is ReLu. The classification layer consists of 3 layers of neurons described as follows: the first layer consists of 100 neurons, the second layer is made up of 20 neurons, all using the read back activation function and finally we have a single neuron in the last layer because it is a binary classification and therefore the most appropriate activation function is sigmoid. During training procss, the error correction function is BinaryCrossentropy.

4.4 Results

The evaluation of our model was carried out according to 04 criteria such as: **Classification rate**(this is the quotient between the number of correct predictions and the total number of predictions), **precision rate**(it allows us to know the percentage of positive predictions that are correctly made), **Recall**(this metric shows the percentage of positives correctly predicted by the model), **F1 score**(This is the harmonic mean of precision and recall). These are summarize in the table bellow:

Table 3. Summarise table of result in our dataset and UCI dataset.

Dataset	Accuracy	Precision	recall	f1 score	support
our dataset	0.995	0.983	0.98	0.981	4124
UCI dataset	0.967	0.945	0.94	0.94	1115

4.5   Discussions

Although the proposed model presents satisfactory results, we note some short-comings or dissatisfactions. Indeed, given the very unbalanced size of the data, we believe that the very high accuracy rate at the level of single messages(Hams) results from the fact that the latter are very similar and of short sizes compared to UCI dataset. In addition, padding the size of each SMS to 144 with zeros impacts the accuracy rate of spam detection. We also believe that the accuracy rate in terms of spam detection still needs to be improved due to the very small number. It is also interesting to say that the dataset compared to UCI dataset seems to be similar for SMS Spam, as the latter is in most cases different on 2 or 3 words which seems to provide us with a biased accuracy rate. Moreover, the fact that most Hams messages are small (less than 50 words) as opposed to the size of Spam messages of 100 to 144 words would be a problem in applying more than 100 filters in the convolution layer. We can also say that the word2vec model does not capture semantic or long-term dependency, which may be a limitation of our model. Moreover, the CNN models with the GlobalMaxpooling function that we use only capture the maximum feature but not the small granularities. It would be better to explore other algorithms such as LSTM and transformer. The latter exploit semantic dependency and also the position of words in a text.

5   Conclusion and Future Work

In this work, we apply the convolutional neural network model on the SMS spam dataset in the Cameroonian context. From the methodology carried out, which was the removal of punctuation, the removal of French and English stops word on the dataset, we can say that this model presents satisfactory results. The evaluation of this model according to the metrics: accuracy, precision, recall, F1-score gave respectively: 99.59%, 98.3%, 98%, 98.1%. Although this model presents better results, we note that it also has limitations in terms of the very unbalanced data size, the increase in the size of each message to 144 by zeros which would bias the accuracy of our model. As a perspective, we intend to explain why explore the contribution of LSTMs on this dataset while trying to exploit the long term dependence. We intend to set up an explicability model, making it possible to justify why this neural model classifies a message as such.

References

1. Suparna. D.G., Soumyabrata. S.: SMS spam detection using machine learning. J. Phys. Conf. Seri. **1797** (2021). https://doi.org/10.1088/1742-6596/1797/1/012017
2. Tiago, A.: SMS spam collection data set from UCI machine learning repository. https://www.kaggle.com/datasets/uciml/sms-spam-collection-dataset
3. Mathappan, N., Rs, S.: SMS Spam Detection using Deep Neural Network. Int. J. Pure Appl. Math. **119**, 2425–2436 (2018)
4. Kumar, P., Prakash, J.: Deep learning to filter SMS spam. Future Gener. Comput. Syst. **102**, 524–533 (2020). https://doi.org/10.1016/j.future.2019.09.001
5. Cormack, G.: Email spam filtering: a systematic review. Found. Trends Inf. Retrieval **1**, 335–455 (2006)
6. Shaaban, M.A.: Deep convolutional forest: a dynamic deep ensemble approach for spam detection in text. Complex Intell. Syst. **8**, 4897–4909 (2022)
7. Ghourabi, A., Mahmood, M.A.: A hybrid CNN-LSTM model for SMS spam detection in Arabic and English messages. Future Internet **12**, 335–455 (2020)
8. Jain, G., Sharma, M., Agarwal, B.: Optimizing semantic LSTM for spam detection. Int. J. Inf. Technol. **11**(2), 239–250 (2018). https://doi.org/10.1007/s41870-018-0157-5
9. Almeida, T., Gomez, H.: Towards SMS spam filtering: results under a new dataset. Int. J. Inf. Secur. Sci. **2**, 1–18 (2013)
10. Yeshwanth, Z.: Spam text classification using LSTM recurrent neural network. Int. J. Emerging Trends Eng. Res. **9**, 1271–1275 (2021)
11. Raj, H., Weihong, Y.: LSTM based short message service (SMS) modeling for spam classification. In: ICMLT '18: Proceedings of the 2018 International Conference on Machine Learning Technologies, pp. 76–80 (2018)
12. Popovac, M., Karanovic, M.: Convolutional neural network based SMS spam detection. In: 2018 26th Telecommunications Forum (TELFOR), pp. 1–4 (2018)
13. Almeida, T.A., Gomez, H.: Contributions to the study of SMS spam filtering: new collection and results. 259–262 (2011)
14. Poomka, P., Pongsena, W.: SMS spam detection based on long short-term memory and gated recurrent unit. Int. J. Feature Comput. Commun. **8**, 11–15 (2019)
15. Porter, G., Hampshire, K.: Mobile phones and education in Sub-Saharan Africa. J. Int. Dev. 28 (2015)
16. Almeida. T.A., Hidalgo, J.M.G.: SMS spam collection (2023). http://www.dt.fee.unicamp.br/~tiago/smsspamcollection/
17. Gupta, M., Bakliwal, A.: A comparative study of spam SMS detection using machine learning classifiers. In: 2018 Eleventh International Conference on Contemporary Computing (IC3), vol. 28, pp. 1–7 (2018)

Time Aware Implicit Social Influence Estimation to Enhance Recommender Systems Performances

Armel Jacques Nzekon Nzeko'o[1,2(✉)], Hamza Adamou[1,2],
Thomas Messi Nguele[1,2,3], and Bleriot Pagnaul Betndam Tchamba[1]

[1] University of Yaounde I, FS, Computer Science Department, Yaoundé, Cameroon
{armel.nzekon,adamou.hamza,thomas.messi,
pagnaul.betndam}@facsciences-uy1.cm
[2] Sorbonne Université, IRD, UMI 209 UMMISCO, 93143 Bondy, France
[3] University of Ebolowa, HITLC, Computer Engineering Department, Ebolowa,
Cameroon

Abstract. Nowadays, e-commerce websites like Amazon, streaming platforms like Netflix and YouTube, and social networks like Facebook and Instagram play a significant role in our daily lives. However, with the constantly growing addition of items on these platforms, it becomes challenging for users to select the products that interest them. Hence, the implementation of recommender systems to facilitate this selection process. To enhance these recommender systems, some studies integrate social influences through trust and friendship information among users to whom recommendations are intended. However, this process of estimating social influence does not consider time and is based on explicits relationships of trust between users, which is not reassuring since these informations are not always available on e-commerce sites. In this paper, we propose to incorporate the temporal aspect into the process of estimating social influences, but using implicit trust informations (ratings that users give to items), which is much more available. Specifically, we have modified the basic recommender systems by incorporating the results of time aware social influence estimations based on implicit trust (through ratings of users). For our experiments, we used the epinions and ciao datasets, which are two platforms where users provide reviews on products from various domains. These experiments demonstrate that considering the temporal aspect of social influence effectively contribute to the improvement of these recommender systems performances. More precisely, we obtained an improvement from 0.902 to 0.833 following the MAE (Mean Absolute Error) metric and from 1.179 to 1.071 following the RMSE (Root Mean Square Error) metric for the Epinions dataset, and for the Ciao dataset, we obtained an improvement from 0.738 to 0.687 following MAE and from 1.023 to 0.953 following RMSE metric.

Keywords: Time aware social influence · Implicit trust ·
Recommender system · Trust-based recommender systems

© The Author(s), under exclusive license to Springer Nature Switzerland AG 2024
P. Melatagia Yonta et al. (Eds.): CRI 2023, CCIS 2085, pp. 15–29, 2024.
https://doi.org/10.1007/978-3-031-63110-8_2

1   Introduction

For the past decade, online platforms have been offering a wide range of items (products, films, social network posts) to their users. For example, on an e-commerce platform like Amazon, products are sold to users; on streaming platforms like Netflix and Youtube, films and series are made available to users; on social networks like Facebook and Instagram, users consult publications.

However, with the ever-increasing number of items made available to users, it is becoming increasingly difficult for a given user to manually browse through all the products on the platform in order to select those that interest him or her. This manual search for products is time-consuming for the user, and can lead to a loss of interest in the platform, resulting in a loss of revenue for the platform's managers. It was with a view to alleviating these problems that recommender systems were set up. Their aim is to offer the user a restricted set of items from among millions that best match their preferences.

Recommender systems are based on several approaches to filtering information, notably collaborative filtering, which assumes that users who have had the same preferences in the past will have the same preferences in the future. It does this by recommending items based on the past behavCiaoior of similar users, correlating users with similar preferences and interests. We also have content-based recommendation systems, which assume that a user who has liked products in the past will like other products in the same category in the future. Finally, we have hybrid recommender systems, which are in fact a combination of the two types of approach mentioned above.

The above approaches have some limitations, such as not consider additional information such as friendship and trust relationships between users. So, in addition to these approaches, we have trust-based recommender systems that use trust informations between users to make recommendations.

Based on trust information between users, it is possible to estimate a rate of this trust, here called social influence. Very few studies on trust-based recommendation systems take into account the temporal aspect of social influence. In other words, they assume that these influences do not change over time, but in reality this is not the case. In this paper, we propose a new way of estimating social influence, taking into account the temporal aspect of this influence.

The rest of the paper is structured as follows: In Sect. 2, we present a state of the art on trust-based recommender systems, in Sect. 3, we present our idea in detail. In Sect. 4, we present the experiments and results obtained, and finally, in Sect. 5, we offer a conclusion and outlook.

2   Literature Review on Trust-Based Recommender Systems

In this section, we present existing work on recommender systems that incorporate information relating to trust and friendly relations between users, more specifically known as "trust-based recommender systems". To carry out this task, we start by presenting in Sect. 2.1 the generalities on recommender systems, through the different approaches to information filtering namely collaborative filtering, content-based filtering and hybrid filtering. In Sect. 2.2, we present recommender systems based on explicit trust, through the computation of trust and the integration of this computation in classical recommender systems. Subsequently, we present recommender systems based on implicit trust in Sect. 2.3, focusing exclusively on how to estimate this implicit trust. Finally, we close the section by outlining some of the limitations of existing work.

2.1 General Information on Recommender Systems

Recommender systems are software tools and techniques that provide suggestions for items that may be useful to a user. Suggestions relate to various decision-making processes, such as items to buy, music to listen to or online news to read [5]. As stated here [11], "the recommendation system helps to cope with information overload and provide personalized recommendations, content and services". We can also define a recommender system as a set of information filtering techniques that predicts the rank or preference a user assigns to an item among a set of similar items (films, music, books, news, images, web pages, etc.) that are likely to be of interest to him or her.

Recommender systems are based on several approaches to information filtering, including collaborative filtering, which assumes that users who have had the same preferences in the past will have the same preferences in the future, content-based filtering, which assumes that a user who has liked items in the past will like other items in the same category in the future, and hybrid filtering, which combines the two types of approach mentioned above. In the following, we'll discuss these different approaches in greater detail.

Collaborative Filtering. The principle of collaborative filtering is to recommend to a user items that other users similar to him or her have enjoyed. It does this by recommending items based on the past behavior of similar users, correlating users with similar preferences and interests. For example, if a user u_1 shares the same preferences as user u_2 and u_1 watches a video on Netflix, and expresses an interest in it, then the video is likely to be recommended to u_2. In this filtering approach, we encounter memory-based techniques like K-nearest neighbors [2], recommendation graphs [1], and machine learning model-based techniques like matrix factorization (MF) [2] and neural networks.

Content-Based Filtering. Content-based recommendation systems assume that a user who has liked items in the past will like other items in the same category in the future. To achieve this, content-based recommendation evaluates the similarity between items based on their characteristics (genre, actors, author), with the aim of recommending new items similar to those previously selected by the target user [13]. This filtering is carried out in three stages: data pre-processing, profile learning (users and items) and recommendation of new items to users.

Hybrid Filtering. When a user watches a film for the first time on a streaming platform, this may be due to the film's attributes. This suggests that integrating content information into collaborative filtering techniques can have a positive impact on recommendation quality. This idea has inspired the design of new types of so-called hybrid recommender systems, in which the aim is to combine the best of both approaches to create even more effective recommender systems.

The predominant collaborative filtering approach remains central to the literature on recommender systems. However, its limitations, notably the cold-start problem when a new user or product is introduced to the platform, as well as the challenges associated with sparse data in the user-item matrix (the matrix of ratings that users assign to items), can hinder the optimization of recommendation performance. Alongside user-item data matrices, information about users' social relationships is often available, and sometimes data clearly indicates which users trust the system, as observed on platforms such as Epinions and Ciao. These observations have prompted research into recommender systems based on explicit trust. This work aims to solve some of the problems inherent in collaborative filtering while exploiting information linked to social relations and trust.

2.2   Recommender Systems Based on Explicit Trust

Explicit trust occurs when trust information between users is available and provided by the users themselves. For example, a user u_1 may openly declare that he trusts another user u_2. To give an overview of this work, this section is divided into two main parts: the first is dedicated to models of trust and the second describes the integration of trust in recommender systems.

Trust Models. In this subsection, we provide a brief overview of trust and the techniques that can be can be used to measure trust between users of social media in general and recommender systems in particular. This information on trust will serve as a basis for improving conventional recommender systems. We begin this section by defining trust and its properties. The section continues with a description of global and local trust. It concludes with a description of how trust values are estimated, based on explicit trust information provided by users.

Definition and properties of trust
    In the social context, Trust generally refers to the fact that one person has faith in the words and actions of another. It extends to relationships within social groups such as family, friends, a community or an organization [14] . In the field of recommender systems, trust is defined in terms of a user's ability to provide relevant recommendations to another user, as specified by Guo et al.   [3]. It should be noted that, in this study, only trust between users in digital platforms is considered, excluding trust such as a user's trust in real life community.

Trust can be measured in two main ways: binary or continuous. Binary trust simplifies the expression by limiting it to two possible states: one user trusts another, and the other does not. Platforms such as Amazon and eBay illustrate this approach using binary trust values (0 and 1). Continuous trust, on the other hand, offers greater finesse by assigning real numbers to represent the degrees of trust relationships between users.

With regard to the properties of trust, we can cite transitivity, asymmetry, context dependence and personalization [3, 6, 9]. These properties are extracted from the definition of trust between people and form the basis for the creation of trust measurement algorithms. Transitivity enables explicit trust to be propagated along the paths of the trust network to reach other users. Trust is asymmetrical, as it is a subjective and personal relationship between users. It is also context-dependent, as trusting someone on one subject does not guarantee trust on other subjects. For example, a user who is trustworthy in technology is not necessarily trustworthy in gastronomy. Finally, trust is personalized, as the degree of trust one person has in another can vary from person to person. This property is used to define and formulate local trust.

Global and local trust measures

Trust is defined as a relationship between two individuals, where the weights associated with trust reflect the degree of credulity one individual accords to another. Trust metrics facilitate the calculation of trust weights between network users. This view situates trust as a local attribute. Beyond this local view, it is possible to conceive of trust as a global measure. In this way, each user is assigned a global value that reflects his or her reliability on the scale of the entire network.

Global trust metrics are simpler and less time-consuming to calculate than local metrics, since local metrics are calculated for each pair of users. However, local trust models can represent a user's interests more accurately than the global approach. Indeed, local trust models bring personalized estimation of trust.

Calculation of trust values

On some digital platforms, users explicitly declare their trust in other users. This can be seen on platforms such as Epinions and Ciao, where users formally express their trust in others. These declarations of trust play a crucial role in improving user recommendations. When a user u_1 declares that he trusts u_2, a value representing u_1's degree of trust in u_2 is calculated. Then, the number of values calculated depends on the available number of explicit trust relationships.

Jian-Ping Mei et al. [2] have proposed two approaches to calculating trust using data from Epinions. In the first approach, the number of items rated by u_2 is used to estimate u_1's trust in u_2. In the second approach, the number of users declaring that they trust u_2 is used to estimate this trust. However, this work has its limitations, notably a method of estimating global trust that does not take into account the personalization of trust. In addition, it neglects the temporal aspect, assuming that trust between two users remains constant over time, whereas in reality it can vary

We can also mention the work of Nzekon et al. [1], where the method consists of assigning the value 1 in the case of an explicit declaration of trust and 0 otherwise. Here too, we note that the estimated trust does not consider time.

Integrating Trust into Recommender Systems. Once explicit trust information has been exploited to estimate implicit trust between users, it becomes essential to integrate it into conventional recommender systems. This makes it possible to create trust-based recommender systems.

Users are more willing to accept recommendations from trusted friends than from strangers, as studies show [12]. Integrating trust into recommender systems improves the relevance of suggestions and the user experience. These trust-based systems use trust and collaborative filtering techniques for more personalized recommendations, overcoming challenges such as cold start and data sparsity. They fall into two categories: memory-based and machine learning model-based.

Collaborative filtering based on memory and trust

An alternative approach to traditional collaborative filtering emphasizes friendly relationships over stranger connections. This method incorporates trust data to reinforce recommendations from trusted users while limiting the impact of others. For example, Mei et al. [2] used trust data from Epinions to improve the K-nearest neighbor (KNN) model. They computed an inter-user trust matrix, then predicted user scores on items based on trusted neighbors. The formula for predicting user u's score on item i is given by the Eq. 1:

$$\hat{r}_{ui} = \mu_u + \frac{\sum_{v \in P_u(i)} Trust(u,v).(r_{vi} - \mu_v)}{\sum_{v \in P_u(i)} |Trust(u,v)|} \tag{1}$$

where $Trust(u,v)$ represents the trust that u places in v, and $P_u(i)$ corresponds to the set of users whom u trusts and who have rated item i.

Model-based and trust-based collaborative filtering

In this category, model-based collaborative filtering techniques, in particular matrix factorization, are widely employed. These approaches are based on the idea that users' preferences are similar to or influenced by those of trusted users. For example, Mei et al. [2] incorporate trust into matrix factorization. The underlying idea is to adjust a user's latent factor (numerical values that describe a user) to bring them closer to those of users he trusts and those who trust him.

Following the analysis of the Sect. 2.2 on explicit trust-based recommendation systems, it is observed that a significant number of works rely on explicit trust relationships declared by users on the platform. However, this data is not commonly available on most digital platforms. The lack of information about explicit trust relationships has given rise to a new line of research, that of recommender systems based on implicit trust. These do not rely on explicit trust information, but exploit other, more readily available data, such as the history of users' actions on items.

2.3   Recommender Systems Based on Implicit Trust

To talk about of recommender systems based on implicit trust, two conditions must be met. Firstly, the trust between users must be estimated, and then this trust must be integrated into a conventional recommender system to produce a trust-based recommender system. The second condition has already been presented in the case of explicit trust. Thus, in this section, we will focus exclusively on the estimation of implicit trust.

In implicit trust models, there is no explicit trust information, and the aim is to use other information to build a network with trust values between users. In [10], Ziegler shows that there is a relationship between similar user preferences and trust between them. This means that people who share the same interests and tastes tend to trust each other more. We can therefore conclude that it is reasonable to use measures of user preference similarity to infer implicit trust values. Most of these techniques are based on the similarity of user profiles and the history of explicit ratings.

- **Similarity of user profiles** : Similarity between two users is defined by whether they are linked in a social network, have friends in common, or like the same items or categories of items.
- **History of explicit ratings** : Similarity is high between two users, if they rate the same items in the same way. Users who assign similar ratings are more likely to trust each other.

Using the criterion of explicit rating history, Nzekon et al. [1] estimated the implicit trust between users of the Epinons and Ciao platforms, which they then integrated into the graphs. To estimate implicit trust, they used the Jaccard similarity presented by the Eq. 2

$$Jaccard(u,v) = \frac{|I_u \cap I_v|}{|I_u \cup I_v|} \qquad (2)$$

with I_u (resp. I_v) the set of items purchased by user u (resp. v).

This way of estimating trust between users does not require explicit trust information. Instead, it relies on the history of explicit user ratings for items, which is reassuring given that this information is more widely available. However, this approach to confidence estimation has a number of limitations. Firstly, it seems to treat trust as a symmetrical relationship, whereas in reality, trust is an asymmetrical relationship. For example, if a user u_1 grants a trust of 0.8 to u_2, this does not necessarily mean that u_2 grants the same trust to u_1. Moreover, this similarity does not take into account the temporal aspect of trust. However, the fact that one user u_1 reproduces the behavior of another user u_2 over time is an indication of u_2's influence on u_1.

In this section on work in the field of recommender systems based on implicit trust, Ziegler's research has highlighted the possibility of estimating implicit trust by exploiting users' rating histories and the similarities between their profiles. The work of Nzekon et al. also explored the estimation of implicit trust using Jaccard similarity. However, a common shortcoming in these studies is the lack of consideration of time in trust estimation, despite the fact that replication of one user's behavior over time is an indication of another user's influence. To address these shortcomings, the next section proposes to incorporate this temporal aspect into trust estimation, adopting the term "social influence" because of the asymmetric nature of the trust reflected by the method.

3 Using Time Aware Implicit Social Influence in Recommender Systems

In this section, we present the use of time in the estimation of social influence between users. We begin by presenting the estimation of social influence with time taken into account in SubSect. 3.1. Finally, in Subsect. 3.2, we show how to integrate these social influences into the K-nearest neighbor model.

3.1 Time Aware Implicit Social Influence Estimation

The approach proposed in this paper is inspired by Jaccard's similarity measure as used by Nzeko'o et al. [1]. This is a symmetrical similarity measure, i.e. the similarity between the user u and v is the same as that between v and u. The symmetry of this similarity makes it unsuitable for estimating influence between users, as social influence is an asymmetrical measure.

Jaccard's similarity is as follows:

$$Jaccard(u, v) = \frac{|I_u \cap I_v|}{|I_u \cup I_v|} \tag{3}$$

The one we propose is as follows:

$$InfSocB(u, v) = \frac{|I_u \cap I_v|}{|I_u|} \tag{4}$$

I_u is the set of items that user u has rated positively. This measure aims to better capture the influence exerted by v on u, by expressing the proportion of items that u appreciates due to v's (the influencer's) prior appreciation of these items. Although the basic idea is based on this principle, it should be noted that a gap arises when v (the influencer) acquires the item after u. In such a scenario, it is irrelevant to conclude that u has been influenced. To remedy this limitation, we have adjusted the formula, incorporating the condition that u must have purchased the item after v. The modified formula 5, presented as the second version of our approach, takes this consideration into account for a more accurate assessment of influence.

$$InfSocS(u, v) = \frac{|u \rightarrow v|}{|I_u|} \tag{5}$$

The arrow pointing to the right expresses the fact that u follows v and therefore that u is influenced by v. The numerator refers to the number of items u bought after v.

Let's consider the case where u buys the product 2 d after v, and let's also consider the case where u buys the product 2 years after v. In reality, the influence that v exerts on u in the first case is much greater than in the second, because the shorter the time between u's purchase of the same product and v, the greater the influence exerted on u. But the second version of our idea doesn't take this into account. For this reason, we propose a third version of our idea, which takes time into account when estimating the influence a user has. The formula is as follows:

$$InfSocT(u,v) = \frac{\sum_{|u \to v|} f(t_u - t_v)}{|I_u|} \tag{6}$$

t_u and t_v are respectively the times at which u and v rated the item, and f is a temporal penalty function. In this work, the idea behind these functions is to give a high weight to influence relations for which the duration between the two purchases (that of u and v) of the same item is small, and to decrease the weight in the opposite case. We have presented here two (02) functions used in [1] in the graphs to penalize the weight of the oldest edges and give a high weight to the most recent edges:

– **Exponential decay function (EDF):** It is illustrated in Fig. 1 and has the expression $f(x) = e^{-x.ln(2)/to}$. *to* is the half-life, i.e. after a time *to* elapsed between the purchase of v and that of u, the weight of the influence of v on u in relation to a given item diminishes by half.
– **Logistic decay function (LDF):** It is illustrated in Fig. 2 and has the expression $f(x) = 1 - 1/(e^{-K(x-to)} + 1)$. K is the slope of the decay curve and *to* the half-life as well. The difference with the exponential function is that the logistic function cancels when $x = 2 * to$.

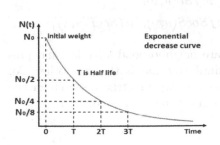

Fig. 1. Exponential decay function (EDF)

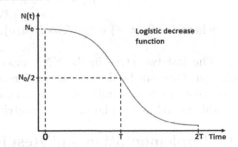

Fig. 2. Logistic decay function (LDF)

3.2 Integrating Social Influence in Recommender Systems

To implement our different ideas through a classical recommendation system, we chose the User K Nearest Neighbors (U-KNN) model. This is a neighborhood-based model which, based on similarities between users, makes product recommendations to a user based on the ratings of that user's neighbors.

The prediction steps of the rating of the user u for the item i are as follows:

– Calculate the similarity between u and all other users in the database
There are two main functions for calculating similarity: **Cosinus** and **Pearson**. The Pearson function is better suited to this model, as it takes into account the fact that users have different ways of rating products.
Let I_u be the set of indices of the items that u has rated and μ_u the average of u's rating. The Pearson correlation coefficient between two users u and v is given by the formula 7.

$$Sim(u,v) = Pearson(u,v) = \frac{\sum_{i \in I_u \cap I_v}(r_{ui} - \mu_u).(r_{vi} - \mu_v)}{\sqrt{\sum_{i \in I_u \cap I_v}(r_{ui} - \mu_u)^2}.\sqrt{\sum_{i \in I_u \cap I_v}(r_{vi} - \mu_v)^2}} \tag{7}$$

where r_{ui} and r_{vi} are respectively ratings that users u and v gave to item i.
– Select the top-k users most similar to u who have already rated the item i (item whose rating we want to predict) :
These users make up the set $P_u(i)$.
– Calculate the prediction $\hat{r}_{ui}$ of the score that user u will assign to item i by Eq. 8

$$\hat{r}_{ui} = \mu_u + \frac{\sum_{v \in P_u(i)} Sim(u,v).(r_{vi} - \mu_v)}{\sum_{v \in P_u(i)}|Sim(u,v)|} \tag{8}$$

To incorporate the social influence previously calculated, we have replaced Pearson similarity $(sim(u,v))$ by social influence $(infSo(u,v))$, and so the score prediction formula becomes :

$$\hat{r}_{ui} = \mu_u + \frac{\sum_{v \in P_u(i)} infSo(u,v).(r_{vi} - \mu_v)}{\sum_{v \in P_u(i)}|infSo(u,v)|} \tag{9}$$

where $infSo(u,v) \in \{InfSocB(u,v), InfSocS(u,v), InfSocT(u,v)\}$

The last two steps in the KNN model are neighborhood selection and prediction. One can be done with a different similarity matrix than the other. So, to explore more possibilities, we take Pearson similarity matrix to search for the neighborhood and social influence matrix to make the score prediction.

4   Implementation and Results

This section is dedicated to the presentation of the experiments carried out and the results obtained. It is structured into three (03) main subsections: the Subsect. 4.1 which sets out the datasets on which the experiments were carried out, the Subsect. 4.2 which details the experimental protocol, and finally the Subsect. 4.3 which presents the results, accompanied by comments.

4.1 Dataset

To carry out the experiments, we made use of publicly available dataset extracts from the Epinions and Ciao platforms [4]. These platforms are dedicated to publishing user reviews of a wide range of items, such as books, DVD, computers, and other items in various categories. Each of these datasets has been modeled as a set of tuples (u, i, c, r, t), meaning that user u has assigned a rating $r \in \{1, 2, 3, 4, 5\}$ to item i at time t, with c representing the category of item i. Both datasets also contain information on explicit trust between users. This data is in the form of tuples (u_1, u_2), meaning that user u_1 trusts user u_2.

Table 1 provides information on these two datasets. minU denotes the minimum number of appearances of a user in the dataset, while minI represents the minimum number of appearances of an item in the dataset. NbNotes quantifies the number of ratings present in the dataset, while NbUsers and NbItems refer, respectively, to the number of users and items present in the dataset.

Table 1. Description des jeux de données Epinions et Ciao.

Name	NbUsers	NbItems	minU	minI	NbNotes	Period
Ciao	889	9053	1	1	12742	2007–2011
Epinions	728	18141	20	2	58717	2006–2010

4.2 Experimental Protocol

In this subsection, we present the segmentation of the dataset, the metrics for evaluating recommendation models and the description of some parameters.

Dataset Segmentation. The task of dividing the dataset must be carried out in such a way as to guarantee the reliability of the resulting recommendation system. To assess this reliability, it is not enough to simply partition the data into training and test sets. It is necessary to implement a mechanism enabling repeated evaluation of the recommender system, thus providing a better insight into its reliability.

In this article, we have therefore opted to use cross-validation with a time window of increasing size. We chose this method because it ensures that all the data in the test set are more recent than those in the training set. This method takes into account the temporal evolution of the data, where the size of the training set increases progressively with time. To implement this method, we define a test window size, which we set at four (04) months, denoted d_{jt}, and optionally an initial training set size, which we set at three (03) years, denoted d_{ja}. Thus, the first [training, test] sample covers the period $[d_{ja}, d_{jt1}]$, the second encompasses the interval $[d_{ja} + d_{jt1}, d_{jt2}]$, and so on, with the final sample covering data from the interval $[d_{ja} + d_{jt1} + ... + d_{jt(k-1)}, d_{jtk}]$. This method, used previously in [7], is illustrated in Fig. 3. In this work, we carried out our experiments using three separate samples.

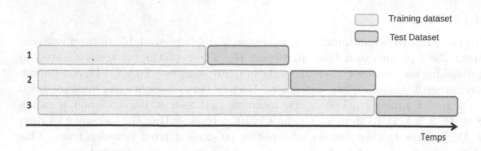

Fig. 3. Cross-validation with increasing time window size.

Evaluation Metrics. As evaluation metrics, we chose two metrics commonly used to assess the quality of score prediction, namely Root Mean Square Error (RMSE) and Mean Absolute Error (MAE). The first corresponds to the square root of the average square score prediction errors, while the second refers to the mean of the absolute error values. The prediction error here is the difference between the actual score and the predicted score. These two metrics are defined by the formulas 10 and 11:

$$RMSE = \sqrt{\frac{\sum_{u,i \in r_{test}} (r_{ui} - \hat{r}_{ui})^2}{|r_{test}|}} \tag{10}$$

$$MAE = \frac{\sum_{u,i \in r_{test}} |r_{ui} - \hat{r}_{ui}|}{|r_{test}|} \tag{11}$$

with r_{test} the test data set.

For each of these metrics, the larger the value, the greater the model error and, consequently, the poorer the model's performance. The fact that the MAE or RMSE decreases indicates an improvement in the performance of the recommender system.

Predefined Parameter Values. To incorporate social influences into the K-nearest neighbor model, we have defined a number of values associated with the model parameters. The description of the parameters and their predefined values is summarized in Table 2.

Table 2. Predefined parameter values for KNN

	Parameter description	Predefined values
K	Number of neighbors	2, 3, 5, 10, 20, 30
To	Half-life of EDF and LDF functions	30, 60, 120, 240, 360 d

4.3   Results and Comments

Before presenting the results, we first present in table 3 the configurations we subjected to experimentation.

Table 3. Description of different configuration codes

Code	Code description
B	Basic model, i.e. no integration of additional information
J	Integrating Jaccard similarity into the model (Eq. 3)
A	Integration of social influence that does not take into account the sequencing of purchases between users (Eq. 4)
S	Integration of social influence, taking into account the sequencing of purchases between users (Eq. 5)
TE	Integration of temporal implicit social influence with exponential decay function EDF (Eq. 6)
TL	Integration of temporal implicit social influence with logistic decay function LDF (Eq. 6)

Figure 4 shows experimental results for the Ciao and Epinions datasets. The "KNN (P-M)" column refers to the KNN model, where Pearson similarity was used to determine the user's neighbors, whatever the configuration. In the "KNN (M-M)" column, on the other hand, the same matrix (representing social influence) was used for both neighbor selection and score prediction. In each cell, the more the value tends towards white, the better the recommender system's performance for that metric. Conversely, the more it tends towards red, the less satisfactory it is. For the Ciao dataset, we note an improvement in both metrics when moving from the basic KNN (B) to the KNN with asymmetric (A) social influence (from **0.738** to **0.718** for MAE and from **1.023** to **0.993** for RMSE). Similarly, when we move from KNN with purchase sequences taken into account (S) to KNN with time taken into account (TL), the results improve (from **0.701** to **0.687** for MAE and from **0.963** to **0.953** for RMSE). Improvements are also noticeable on the Epinions dataset.

Comparing ourselves to the work of Nzekon et al., who estimated implicit confidence using Jaccard similarity (J), we go from **0.719** to **0.687** following the MAE and from **0.993** to **0.953** following the RMSE for the Ciao dataset. For Epinions, we go from **0.845** to **0.833** according to the MAE and from **1.099** to **1.071** according to the RMSE.

Overall, we can say that taking time into account in the estimation of social influence has a positive impact. Not only does it improve the basic K-nearest-neighbor model, but it also improves the same model with the integration of Jaccard similarity. The logistic decay function (LDF) outperforms the exponential (EDF), cancelling out at $2 * t0$. This cancellation expresses the inevitable decrease of influence between users about an item over time, an advantage that EDF does not have. After analysis, the optimal value for to is 30 d.

CIAO	KNN (P-M)		KNN (M-M)	
	MAE	RMSE	MAE	RMSE
B	0.7383	1.0239	0.7383	1.0239
J	0.7197	0.9934	0.7218	0.9948
A	0.7189	0.9933	0.7202	0.995
S	0.7014	0.9631	0.7036	0.9657
TE	0.6994	0.9609	0.7032	0.9656
TL	0.6874	0.9538	0.6954	0.9661

EPINIONS	KNN (P-M)		KNN (M-M)	
	MAE	RMSE	MAE	RMSE
B	0.9028	1.1796	0.9028	1.1796
J	0.8459	1.0999	0.8462	1.1001
A	0.8483	1.1032	0.8485	1.1032
S	0.8476	1.0995	0.8478	1.0995
TE	0.848	1.1002	0.8482	1.1002
TL	0.8334	1.0718	0.8336	1.0724

Fig. 4. KNN results with Ciao and Epinions

5 Conclusion

In this paper, the aim was to estimate social influence from implicit user data (ratings). This eliminated the need for explicit trust relationships between users. Secondly, in estimating this social influence, it was necessary to take into account the temporal aspect of the influence. Experimental results show that taking time into account improves recommendations. we obtained an improvement from 0.738 to 0.687 following the MAE metric and from 1.023 to 0.953 following the RMSE metric for the Ciao dataset, and for the Epinions dataset, we obtained an improvement from 0.902 to 0.833 following the MAE metric and from 1.179 to 1.071 following the RMSE metric. Looking ahead, we'll be carrying out further experiments on several datasets. In addition to this, we will also take into account item categories when estimating social influence, as the influence of one user by another can be high for a given item category and low for another. We also plan to integrate these estimated social influences into other recommender systems such as matrix factorization and recommendation graphs.

Acknowledgments. The success of this research project was made possible the support of all the members of our working team group, specifically the HIgh PERformance DAta Science (HIPERDAS), who actively participated in stimulating exchanges and shared valuable ideas. In addition, the collaborative environment established within our university community played a fundamental role in the progress of this research. In conclusion, we would like to sincerely thank all those who, directly or indirectly, contributed to the success of this work.

References

1. Nzekon, A.J.N.: A general graph-based framework for top-N recommendation using content, temporal and trust information. J. Interdisc. Methodol. Issues Sci. 5 (2019)
2. Jian-Ping, M.: A social influence based trust model for recommender systems. Intell. Data Anal. **21**(2), 263–277 (2017)

3. Guo, G., Zhang, J., Thalmann, D., Basu, A., Yorke-Smith, N.: From Ratings To Trust: an empirical study of implicit trust in recommender systems. In: Proceedings of the 29th Annual ACM Symposium on Applied Computing, pp. 248–253 (2014)
4. Tang, J., Gao, H., Liu, H.: mTrust: discerning multi-faceted trust in a connected world. In: Proceedings of the Fifth ACM International Conference on Web Search and Data Mining, pp. 93–102 (2012)
5. Ricci, F., Rokach, L., Shapira, B.: Introduction to recommender systems handbook. In: Ricci, F., Rokach, L., Shapira, B., Kantor, P.B. (eds.) Recommender Systems Handbook, pp. 1–35. Springer, Boston, MA (2011). https://doi.org/10.1007/978-0-387-85820-3_1
6. Castelfranchi, C., Falcone, R.: Trust Theory: a socio-cognitive and computational model. John Wiley & Sons (2010)
7. Lathia, N., Hailes, S., Capra, L.: Temporal collaborative filtering with adaptive neighbourhoods. In: Proceedings of the 32nd International ACM SIGIR Conference on Research and Development in Information Retrieval, pp. 796–797 (2009)
8. Koren, Y.: Factorization meets the neighborhood: a multifaceted collaborative filtering model. In: Proceedings of the 14th ACM SIGKDD International Conference on Knowledge Discovery and Data Mining, pp. 426–434 (2008)
9. Golbeck, J.A.: Computing and applying trust in web-based social networks. University of Maryland, College Park (2005)
10. Cai-Nicolas, Z., Jennifer, G.: Investigating correlations of trust and interest similarity-do birds of a feather really flock together. Decis. Support Syst. **42**(3), 1111–1136 (2005)
11. Gediminas, A.: Toward the next generation of recommender systems: a survey of the state-of-the-art and possible extensions. IEEE Trans. Knowl. Data Eng. **17**(6), 734–749 (2005)
12. Sinha, R.R., Swearingen, K.: Comparing recommendations made by online systems and friends. DELOS **106**(1), 1–6 (2001)
13. Mooney, R.J., Roy, L.: Content-based book recommending using learning for text categorization. In: Proceedings of the Fifth ACM Conference on Digital libraries, pp. 195–204 (2000)
14. Lewis, J.D.: Trust as a social reality. Soc. Forces **63**(4), 967–985 (1985)

Analysis of COVID-19 Coughs: From the Mildest to the Most Severe Form, a Realistic Classification Using Deep Learning

Fabien Mouomene Moffo[1]([✉]) [iD], Auguste Vigny Noumsi Woguia[1] [iD],
Samuel Bowong Tsakou[1] [iD], and Joseph Mvogo Ngono[2] [iD]

[1] Faculty of Science, Douala University, Douala, Cameroon
fmouomene@yahoo.fr
[2] Douala University, Douala, Littoral, Cameroon

Abstract. Cough is the most recurrent symptom of lung disease. In addition, COVID-19 is an unparalleled lung disease. The spread of this pandemic has resulted in over 600 million positive cases and over 6 million deaths worldwide. Therefore, an efficient, inexpensive, and ubiquitous diagnostic tool is essential to help fight lung disease and the COVID-19 crisis. Deep learning and machine learning algorithms can be used to analyze the cough sounds of infected patients and infer predictions. We made use of constructivist logic. The cough data are from our research lab and the COUGHVID research lab. This Diagnostic approach, based on deep learning and feature extraction from Mel spectrograms, can recognize cough sounds from sick people without COVID-19, with severe and mild COVID-19, and also recognize cough sounds from healthy patients. The model used is a variant of ConvNet. The quiet environments, which allowed the acquisition of data, reduce systematic and random errors in the quality of the audio. The architecture of the convolutional neural networks exploited, gives an overall Accuracy of 90.33%. This system could have a significant positive social impact by minimizing transmission of the virus, speeding up patient treatment, and freeing up hospital resources. In addition, early diagnosis of COVID-19 may also prevent further disease progression and improve the effectiveness of screening efforts.

Keywords: COVID-19 · Diagnosis · Cough · Deep learning · ConvNet

1 Introduction

Cough is the most recurrent symptom of lung disease. In addition, COVID-19 is an unparalleled lung disease, and the spread of this pandemic has resulted in more than 600 million positive cases and over 6 million deaths worldwide [1,2]. The viruses responsible for respiratory diseases are present in the respiratory secretions of patients, which can be projected as droplets around the

P. Melatagia Yonta et al. (Eds.): CRI 2023, CCIS 2085, pp. 30–40, 2024.
https://doi.org/10.1007/978-3-031-63110-8_3

patient during coughing [3]. These droplets are the main sources of contamination. COVID-19 contamination can be direct and indirect [4]. Direct during the patient's cough and indirect after the patient's cough through infected surfaces [5]. Medical and clinical solutions have been used to stop the spread of the disease, such as nucleic acid testing (RNA) and amalgam, polymerase chain reaction (PCR) testing. The diagnosis of COVID-19 is confirmed by RT-PCR testing of nasopharyngeal and oropharyngeal swabs. Test results are typically available within four hours. The samples collected for the initial detection of COVID-19 include nasopharyngeal swabs, lower respiratory tract swabs (sputum, BAL) in cases with parenchymal involvement, and blood samples [6]. The interpretation of RT-PCR test has more weight because of the specificity of the test and sensitivity Segal. Other scientific methods in epidemiology are published for the prevention of COVID-19 [7,8].

Despite the reliability, RT- PCR tests for SARS-CoV-2 pose some problems in use. Requirement of health care personnel, a risk of contamination during collection, the high cost of testing due to the need for expensive reagents and tools [9]. In addition, the processing of the collected samples is not in real-time. These problems, so unpleasant, analogy, prevent RT-PCR tests from being large-scale screening tools. Although the vaccination rate, in order to achieve herd immunity, is high in some countries, it is still very low in some continents affected by the COVID-19 pandemic due to the backlog and the failure of vaccines. In the face of this controversy, experts are uncertain about the achievement of herd immunity, especially with the unpredictability of viral variants. Therefore, an effective, inexpensive, and ubiquitous diagnostic tool is essential to address the lung disease and COVID-19 crisis.

One of the most common symptoms of COVID-19 is a dry cough, which is present in approximately 67.7% of cases [10]. Convolutional neural networks, which are a revolution in artificial intelligence (AI), show promise in enabling the creation of such a solution. Deep learning and machine learning algorithms could be used to analyze the coughing sounds of infected patients and derive predictions. Research laboratories have had to collect sound recordings for COVID-19 patients of all ages, in changing environments, symptomatic or asymptomatic. This data provides an opportunity for AI algorithms to learn pandemic-specific audio patterns for patients. Commonly collected auditory recordings to detect COVID-19 are coughs. Groups such as COUGHVID and COSWARA [11], collect data from cough recordings. Nevertheless, the recorded data is not always objective in terms of sound quality and formats, depending on the tool used and the environment. Missing donor data makes some of the metadata unsuitable for learning algorithms.

We propose a deep learning network based on cough sounds. Our model, which is a variant of ConVNet, successfully learns from cough audio submissions. The model classifies into four categories: severe and non-severe COVID-19 positive donors, healthy patients, and other patients suffering from lung diseases other than COVID-19.

2   Materiel and Methods

We conducted cross-sectional research in a descriptive, exploratory and diagnostic study. Our work has implications for the international population. This research originated in our laboratory before the COVID-19 pandemic. In collaboration with health care workers, we recorded cough samples from healthy patients and patients with respiratory diseases. Data for patients who tested positive for COVID-19 were later completed. Participants were recorded in a quiet environment [12]. Recordings with high ambient noise were excluded. Patients consented to participate, freely and with team probity, in these sound recordings. Since lung diseases and COVID-19 were not exclusively relevant to one country or continent, so we wanted our subjects to be geographically dispersed all over the world. Figure 1 presents the methodology of this study.

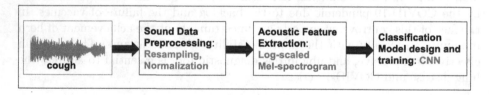

Fig. 1. Methodology of this Study.

2.1   Data Collection

The data analyzed correspond to the audio recordings of coughing patients. These audio data, collected randomly, were recorded from the microphones of cell phones and computers. Our dataset, consisting of 3300 data, describes coughs of healthy patients, sick patients with negative COVID-19 test and patients with severe and not severe COVID-19. Samples from 258 healthy patients and sick patients with a negative COVID-19 test were recorded before the COVID-19 pandemic [13]. The remaining data during the pandemic through the COUGHVID laboratory are public data. The metadata, of the known patient population, indicates the patient's condition, age, gender, and geographic location. Recurrent symptoms in COVID-19 are: cough, fever, and fatigue. High fever is the symptom that has been taken into account to most describe severe COVID-19 [14]. The in vivo severity of covid is due to the presence of hemoptysis, decreased white blood cells and renal failure [15,16]. Figure 2 illustrates the geographical sources of all collected cough sounds and Fig. 3 displays the clinical forms.

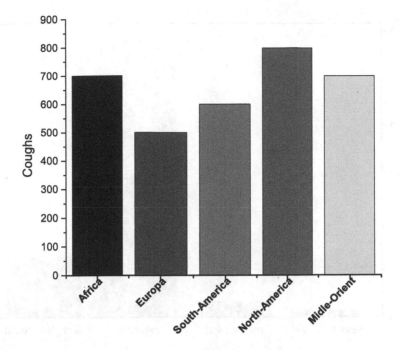

Fig. 2. The Geographical Sources of Dataset.

2.2 Preprocessing

First, the audio signal, like any other incoming signal, must be digitized into a sequence of numbers, called samples, each indicating the air pressure on the microphone at a given time. The raw data collected was in Ogg,Webm and Mp3 formats. We first developed an algorithm for automatic conversion of the data set from each audio format to wav format. The Wav format, unlike the Ogg and Webm formats, offers the best listening quality and has the advantage of not compressing the audio file. These data are grouped into four classes: samples from sick patients without COVID-19 (sick), samples from healthy patients (healthy), samples from people with severe COVID-19 and samples from people with non-severe COVID-19. Samples with different frequencies were normalized to a frequency of 22 khz. A second algorithm was developed to slice the audio samples to have an identical duration of 3 s for the 2D CNN input. Typically, a speech signal requires 10,000 of these samples per second. Figure 4 shows the nature of the 1 D signal for each cough category.

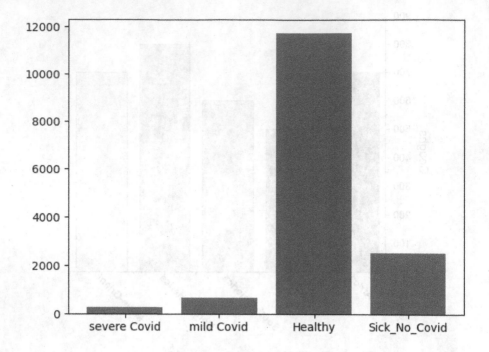

Fig. 3. Clinical forms.

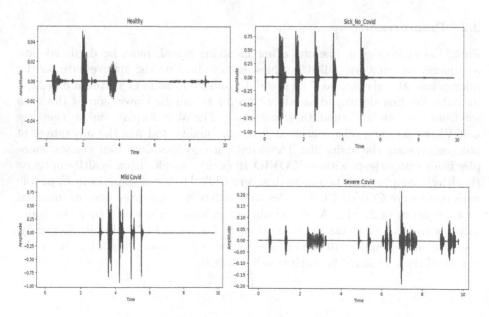

Fig. 4. Signal for each Cough Category.

From the audio data, acoustic features were extracted by the logarithmic scaled mixture spectrogram method. It is a method that allows the representation of features of sound events and the distinction between different types of sound events. Logarithmic spectrograms can provide rich acoustic information from a short recording of audio data. As a result, sound events that occur in a short period of time can be detected efficiently. In order to derive the log-scale spectrogram of the original sounds, three steps are required. first, converting the sound clips from the time domain to the frequency domain by applying the short-time Fourier transform [17]. Then the use of a Hann window of 23 ms with a jump length of 256. Then, the transformed audio sequence was converted to 2D as a spectrogram with 120 graduated frequency bands and magnitudes showing the characteristics of the human audible range, including the decibel scale. A speech signal is thus represented by a spectrogram, a sequence of vectors of dimensions 120×40, with one vector every 10 milliseconds, i.e. 100 vectors per second [18]. The convolutional network takes a window of 120 vectors (its input "image" is therefore 120×40 "pixels") representing cough, and classifies the elementary sound present in the middle of the window [19]. Finally, the derivative of the logarithmically scaled mixed spectrogram was used as an addition of extra channels for understanding and distinguishing acoustic features more efficiently.

2.3 Architecture

Deep learning consists in: building the architecture of a multilayer network by arranging and connecting modules; training this architecture by gradient descent after computing this gradient by backpropagation [20]. Our deep learning architecture is based on convolutional feed-forward neural networks, which uses all connected layers or ConvNet. It takes 3 convolutional layers 3×3 + max-pooling 4×2. Two 1024×1024 fully connected layers in the last two layers and 1024 fully connected layer in the last layer, Dropout applied on the fully connected layers. The 3 layers are connected as described in Fig. 5. The feature map produced by the convolution has positive and negative values, because the weights can be negative. When it passes through the ReLU layer, the ReLU sets the negative values to zero, and leaves the positive values unchanged. The output arrays from the ReLU layer are also called feature maps. This non-linear operation allows the system to detect patterns in the image. The ReLU layer is followed by a pooling layer, the feature map output of the ReLU layer is divided into windows, or rather into non-overlapping tiles. Each neuron in the pooling layer takes one of these windows and calculates its maximum value. In other words, a window has 8 numbers, and the neuron produces the largest of these 8 numbers on its output. This is called max pooling. Pooling is used to produce a representation that is invariant to small shifts in the patterns in the input image. The max pooling produces the largest value in its input, which corresponds to the most marked pattern in its receiver field. the convolutional network is made of a stack of layers of convolutions, ReLU and pooling.

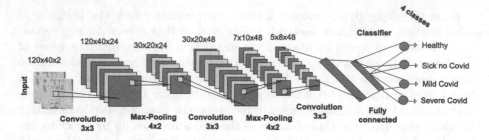

Fig. 5. Architecture.

2.4   Training and Build

Learning is the same as making the adjustment. It consists in adapting the system progressively to the classification by reducing the errors that this system makes. The reduction of errors is minimizing the cost function. To train our model, we use a split method, first separating the dependent and independent features. Then we gave labels to the four classes by coding them from number 1 to 4. The data were split into training and test sets in an 80–20 ratio. We used: the cross entropy for the loss function, the precision score for the accuracy measures and the Adam method as a learning optimizer. The adam method is based on gradient backpropagation [21]. This backpropagation of gradients is the adjoint state applied to multi-layer networks, it allows to calculate the gradient of a cost function. The principle consists in propagating a signal backwards in the network, thus propagating the partial derivatives (gradients). We trained the model for 100 epochs and a batch size of 30 in order to modify the weights or variables of the cost function. The control of the time was done by the recall.

3   Results and Discussions

3.1   Results

The sample studied is a short sequence or excerpt from a human voice recording. These recordings are coughing sounds of patients. They are classified in four categories: Healthy, Sick without COVID-19, Sick with severe COVID-19 and not severe. The quiet environments, which allowed the acquisition of the data, reduce systematic and random errors.The statistical measures used to estimate the numerical results of the algorithms were: Sensitivity (or recall), which measures correctly identified true positives; Specificity (or selectivity, precision), which measures correctly identified true negatives; Accuracy, which measures the total number of samples correctly classified; F1 score, which is the harmonic mean of precision and recall. The architecture of the convolutional neural networks exploited, gives an overall accuracy of the cross-validation using the cough data of 90.33%. Figure 6 displays the outcomes of the feature extraction process employing the MFCC method, while Fig. 7 illustrates the results of the model compilation.

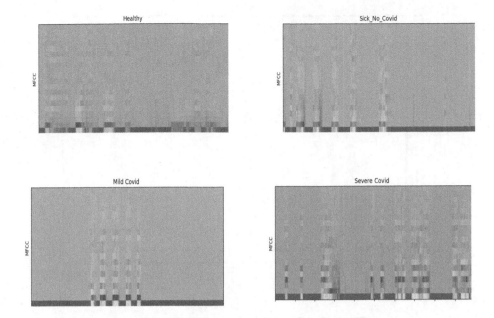

Fig. 6. Results of Feature Extraction.

3.2 Discussions

Our study presents three main results. The first main result is the classification of four categories of cough noises: cough of sick patients without COVID-19, cough of healthy patients and cough of sick patients with severe and not severe COVID-19. The second result concerns the learning by Adam's method and the third result is the extraction of the characteristics of each cough category. It is remarkable to mention that our CNN model with 90.33% accuracy surpasses the models of R. Islam whose CNN model for cough detection had an accuracy of 88.5% and this model also surpasses the work of C. Brown et al. on COVID-19 detection with an accuracy of 80% [11, 22]. Table 1 shows the positioning of the paper in relation to other papers in the field.

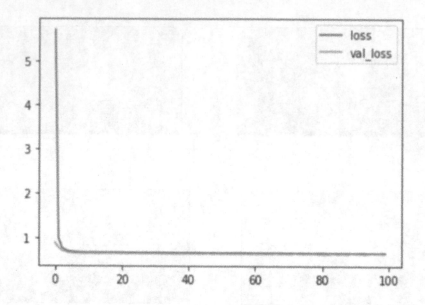

Fig. 7. Results of the Model Compilation.

Table 1. Work positioning.

references	methods	metrics
Yuh-Shyan Chen et al. [23]	method of MFCCs, The Cube-Transformer model	accuracy by 1.5%
MOSCHOVIS et al. [24]	ResAppDx algorithm	76.5% sensitivity
This Work	method of MFCCs, ConvNet	90.33% accuracy %
BROWN et al. [25]	binary machine learning classifier	AUC of above 80%

4   Conclusion

This research described the realization of an intelligent tool for the diagnosis of respiratory diseases based on cough sounds, such as COVID-19. Aid diagnosis approach based on deep learning allows to detect cough sounds of sick people without COVID-19, with severe and mild COVID-19, and also to detect cough sounds of healthy patients. To achieve this, a CNN model was built, trained on data collected by our team, also by other research teams. The experimental result, after validation of the model on data, shows that this CNN model detects with great skill the four classes. Since the experimental report gives an Accurancy of 90.33%. Given the benefit of this proposed model, this work makes an important scientific contribution. Firstly, it allows a quick, vocal diagnosis of respiratory diseases. Secondly, the model distinguishes between mild and severe COVID-19 for an efficient management of patients in hospitals.

Acknowledgments. Our prospective virtue comes from the Lord Jesus Christ. We thank the patients for their agreement to the audio recordings and also the research laboratory COUGHVID for their contribution in data.

References

1. Google News Homepage. https://news.google.com/covid19/map?hl=enUSmid=%2Fm%2F02j71gl=USceid=US%3Aen.. Accessed 25 Jul 2023
2. Social Stigma associated with COVID-19. https://www.unicef.org/sites/default/files/2020-03/Social%20stigma%20associated%20with%20the%20coronavirus%20disease%202019%20%28COVID-19%29.pdf. Updated 24 Feb 2020
3. Kampf, G., Todt, D., Pfaender, S., Steinmann, E.: Persistence of coronaviruses on inanimate surfaces and their inactivation with biocidal agents. J. Hosp. Infect. **104**(3), 246–251 (2020)
4. Chowdhury, R., et al.: Dynamic interventions to control COVID-19 pandemic: a multivariate prediction modelling study comparing 16 worldwide countries. Eur. J. Epidemiol. **35**(5), 389–399 (2020). https://doi.org/10.1007/s10654-020-00649-w
5. Wathore, R., Gupta, A., Bherwani, H., Labhasetwar, N.: Understanding air and water borne transmission and survival of coronavirus: insights and way forward for SARS-CoV-2. Sci. Total Environ. **749**, 141486 (2020). ISSN 0048-9697. https://doi.org/10.1016/j.scitotenv.2020.141486, https://www.sciencedirect.com/science/article/pii/S0048969720350154
6. Biosafety in Microbiological and Biomedical Laboratories, 5th Edition, CDC. Laboratory testing for middle East respiratory syndrome coronavirus, Interim guidanc. (revised) (2018). WHO/MERS/LAB/15.1/Rev1/2018
7. Ranjan, R.: Predictions for COVID-19 outbreak in India using Epidemiological models (2020)
8. Eikenberry, S.E.: To mask or not to mask: modeling the potential for face mask use by the general public to curtail the COVID-19 pandemic. Arizona State University, School of Mathematical and Statistical Sciences, Tempe, AZ, USA, April 8 (2020)
9. Gevertz, J., Greene, J., Sanchez Tapia, C.H., Sontag, E. D.: A novel COVID-19 epidemiological model with explicit susceptible and asymptomatic isolation compartments reveals unexpected consequences of timing social distancing (2020). https://doi.org/10.1101/2020.05.11.20098335, medRxiv
10. Fenaux, H.: Interpretation of single target positivity among SARS-CoV-2 RTPCR result tests. J. Clin. Virology Plus **1**(12), 100021 (2021). ISSN 2667-0380. https://doi.org/10.1016/j.jcvp.2021.100021
11. Muguli, A, et al.: DiCOVA Challenge: dataset, task, and baseline system for COVID-19 diagnosis using acoustics. Proc. Interspeech (2021)
12. Iqbal, T., Kong, Q., Plumbley, M.D, Wang, W.: General-purpose audio tagging from noisy labels using convolutional neural networks. In: Proceedings of the Detection and Classification of Acoustic Scenes and Events 2018 Workshop (DCASE2018), pp. 212–216 (2018). Tampere University of Technology
13. Cases of coronavirus in cameroon. Visited 17 May 2020
14. Chan, J.W.M., Ng, C.K., Chan, Y.H., et al.: Short term outcome and risk factors for adverse clinical outcomes in adults with severe acute respiratory syndrome (SARS). Thorax **58**(8), 686 80 (2003)
15. Gorbalenya, A. E., Baker, S. C., Baric, R. S., et al.: Severe acute respiratory syndrome-related coronavirus: the species and its viruses? A Statement of the Coronavirus Study Group (2020)

16. Verity, R., Okell, L.C., Dorigatti, I., et al.: Estimates of the severity of coronavirus disease 2019: a model-based analysis. Lancet Infect. Dis. **20**(6), 669–677 (2020)
17. Choi, K., Fazekas, G., Sandler, M., Cho, K.: A comparison of audio signal preprocessing methods for deep neural networks on music tagging. In: 2018 26th European Signal Processing Conference (EUSIPCO), Rome, Italy, pp. 1870–1874. IEEE (2018). https://doi.org/10.23919/EUSIPCO.2018.8553106
18. Spanias, A., Painter, T., Atti, V.: Audio Signal Processing and Coding. John Wiley Sons (2006)
19. Sercu, T., Puhrsch, C., Kingsbury, B., LeCun, Y.: Very deep multilingual convolutional networks for LVCSR. IEEE Int. Conf. Acoust. Speech Signal Process. (ICASSP), 4955–4959 (2016)
20. LeCun, Y., Bengio, Y., Hinton, G.: Deep learning. Nature **521**, 436–444 (2015)
21. Farabet, C., Couprie, C., Najman, L., LeCun, Y.: Learning hierarchical features for scene labeling. IEEE Trans. Pattern Anal. Mach. Intell. **8**(35), 1915–1929 (2013)
22. Kamble, M.R., et al.: PANACEA cough sound-based diagnosis of COVID-19 for the DiCOVA 2021 Challenge. Interspeech (2021)
23. Chen, Y.S., Hsu, C.S., Yang, B.X.: The design of a cough disease classification system using cube-transformer. In: 2023 International Conference on Smart Applications, Communications and Networking (SmartNets), pp. 1–6 (2023)
24. Moschovis, P.P., Sampayo, E. M., Porter, P., et al.: A cough analysis smartphone application for diagnosis of acute respiratory illnesses in children. In : A27. Pediatric Lung Infection and Critical Care Around the World. American Thoracic Society, pp. A1181–A1181 (2019)
25. Brown, C., Chauhan, J., Grammenos, A., et al.: Exploring automatic diagnosis of COVID-19 from crowdsourced respiratory sound data. In : Proceedings of the 26th ACM SIGKDD International Conference on Knowledge Discovery and Data Mining, pp. 3474–3484 (2020)

SLCDeepETC: An On-Demand Analysis Ready Data Pipeline on Sentinel-1 Single Look Complex for Deep Learning

Kemche Ghomsi Adrien Arnaud$^{(\boxtimes)}$ (ID), Mvogo Ngono Joseph (ID),
Bowong Tsakou Samuel (ID), and Noumsi Woguia Auguste Vigny (ID)

Laboratory of Applied Computer Science, University of Douala, 2701 Douala,
Cameroon
adrien.ghomsi@weloobe.com

Abstract. SLCDeepETC is an on-demand Analysis Ready Data pipeline designed to automate data processing and cross-platform delivery of interferometry data on Single Look Complex products from Sentinel-1 to predict some environmental phenomena using Deep Learning. By retaining all original sensor measurements, it has been proven that interferometry data on Single Look Complex products, when analyzed with Deep Learning, can better inform data restoration, coherence estimation, classification, and automatic target recognition. However, the large data volume, interferometry processing complexity, interferometry data interpretation difficulties, and heterogeneous framework codebases for Deep Learning algorithms pose challenges to Deep Learning model training, hindering radar interferometry domain research. Through the ETC (Extract, Transform, and Cross-platform delivery) pipeline mechanism, SLCDeepETC preserves essential details and achieves significant speedups and superior performance for radar interferometry analysis with Deep Learning on real-world raw Sentinel-1 Single Look Complex data products. The proposed SLCDeepETC pipeline, on demand, can generate large Analysis Ready Datasets of interferometry derivative time series data on Single Look Complex products for Deep Learning model training through inter-framework codebase functionality, accelerating Deep Learning research in the radar interferometry domain.

Keywords: SAR · Sentinel-1A/1B · Radar Interferometry · Single Look Complex Products · Data Processing · Deep Learning · Cross-Platform Delivery · On-Demand Analysis Ready Data · Training Datasets Generator · Real-World Data

1 Introduction

Deep Learning (DL) is an increasing trend in big data analysis, and there is a growing amount of raw Sentinel-1 open data available. In the Sentinel-1A/B system (S-1), the SAR sensor beams millions of radar signals back to Earth in the

© The Author(s), under exclusive license to Springer Nature Switzerland AG 2024
P. Melatagia Yonta et al. (Eds.): CRI 2023, CCIS 2085, pp. 41–52, 2024.
https://doi.org/10.1007/978-3-031-63110-8_4

form of microwaves, generating Single Look Complex (SLC) data, where each pixel is a complex number with amplitude and phase values. SLC products are the best source for SAR image analysis, although SLC products and exploitation software is difficult to use without experience. DL has shown promise in remote sensing applications [12, 27], including image preprocessing [7], change detection [22], accuracy assessment [25], and classification [14]. However, the lack of available interferometry Analysis Ready Datasets (ARD) on raw Sentinel-1 SLC products and the heterogeneous framework codebases of Deep Learning algorithms present a major challenge in training DL models on interferometry data for environmental applications. This handicap restricts the discovery and exploitation of the full information potential of the radar interferometry data's black box, which can upgrade and enhance the capabilities of radar interferometry processing and domain applications. This paper addresses this challenge by providing the SLCDeepETC pipeline, an on-demand "extract, transform, and cross-platform delivery" of interferometry time series data on real-world raw Sentinel-1 SLC products as Analysis Ready Dataset (ARD) of Differential Interferometric Synthetic Aperture Radar (DInSAR) pool products generators dedicated to DL models training for predicting some environmental phenomena.

Materials and Methods - Sect. 2 - introduce the problem and related work on SAR Sentinel-1 pipelines. Section 3 - presents the proposed SLCDeepETC pipeline, including results and their benefits and limitations.

2 Materials and Methods

This section discusses three key challenges when using deep learning for interferometry data mining on Single Look Complex products. It then presents the dataset provider, the Copernicus open access hub. Finally, it describes the ETC Pipeline Approach for Deep Learning and how the proposed pipeline aims to address these challenges.

2.1 Description of the Problem

SLC level-1 dual-polarized products, in its raw form, are largely interpretive in nature and need to be transformed into a quantitative tool to be useful in predicting phenomena with deep learning. However, obtaining quantitative interferometric amplitude, coherence, and phase values is not sufficient to train DL models. The amplitude values vary widely, which affects the effectiveness of deep learning algorithms. To avoid poorly controlled variance, all amplitude values need to lie between 0 and 1. Secondly, phase values belonging to $[-\pi, \pi]$ cannot be used in deep learning processing algorithms due to the phenomenon of branch cuts, which represent high-frequency motion signals. To preserve all the information contained in the phase values, the phase values need to be normalized. Lastly, the real-world raw SLC data is an Xarray tensor, which is not compatible with most deep learning libraries designed for tabular data. Interferometric Xarray tensors need to be adapted for deep learning frameworks with abstract codebase.

2.2 Related Work

The use of deep learning for mining interferometric data on Single Look Complex products presents three key challenges. Several tools exist, including SenSARP [23], snapista [15], pyroSAR [20], S1TILING, Diapason [13], and DiapOTB [5], which offer a wide range of features for preprocessing and generating interferometric time series. These tools provide advanced features for working with complex SAR data, but their complexity can make them difficult for inexperienced users to use, often requiring a steep learning curve. Additionally, integrating these tools with Deep Learning pipelines can be complex and require additional effort. In comparison (Table 1), the SLCDeepETC solution stands out for its ability to simplify SAR data preparation for deep learning, seamless integration with Deep Learning pipelines, and providing a comprehensive solution for generating analysis-ready datasets for deep learning. Thus, while the mentioned tools offer powerful features for handling SAR data, the SLCDeepETC solution stands out for its ease of use and its ability to simplify the process of generating data ready for Deep Learning.

Table 1. Comparison of our pipeline technique with state of art.

	Operational level			Techniques		Evaluation criteria[a]				
Pipeline	E	T	C	DeepINSAR	InSAR	Acc.	M.P.	P.T.	C.C.	Scal.
Our solution	✓	✓	✓	✓	✓	✓	✗	✓	✓	✓
SenSARP	✗	✓	✗	✗	✓	✓	✗	✓	✓	✗
Sanpista	✗	✓	✗	✗	✓	✓	✗	✓	✓	✗
PyroSAR	✗	✓	✗	✗	✓	✓	✗	✓	✓	✗
Diapason	✓	✓	✗	✗	✓	✓	✗	✓	✓	✗
S1Tiling	✓	✗	✗	✗	✓	✓	✗	✓	✓	✗

Legend: 1. E: Extract, 2. T: Transform, 3. C: Cross platform delivery, 4. DeepIN-SAR: InSAR techniques applications for Deep learning, 5. InSAR: For InSAR standard techniques applications, 6. Acc. - Accuracy, 7. M.P. - Model Performance, 8. P.T. - Processing Time, 9. C.C. - Computational Complexity, 10. Scal. - Scalability.
[a] **Acc.** Measure the accuracy of the pipeline in generating analysis-ready datasets for deep learning from raw data; **M.P.** Evaluate the performance of deep learning models trained using the datasets generated by the pipeline; **P.T.** Analyze the time required for the pipeline to transform raw data into datasets ready for deep learning, focusing on the efficiency of the process; **C.C.** Assess the computational complexity of the pipeline to provide an indication of its resource requirements; **Scal.** Study the pipeline's capability to be extended to different platforms, data types, and application scenarios.

2.3   Dataset

This section presents the data provider for the SLCDeepETC pipeline, highlighting the Single Look Complex Level-1 dual-polarized products from Sentinel-1.

Copernicus Open Access Hub. Since 2011, the Copernicus program [6] has promoted free and open access to the observation data of Earth via satellite and other sources, creating new opportunities for scientific and industrial communities. Sentinel-1 (S-1) is an imagery radar mission that provides continuous images, even in low light and bad weather conditions. SAR designates a synthetic aperture radar, which is an imaging radar processing received data to improve resolution in azimuth. S-1 allows for the acquisition of images in different modes, including strip map, interferometric wide swath, extra-wide swath, and wave modes, and provides the SAR products for each mode at three levels: level-0 (unfocused SAR raw data), level-1 (SLC and Ground Range Detected - GRD data), and level-2 (Ocean geophysical product derived from level-1). We used VV + VH dual-polarized SLC (level-1) images acquired in interferometric wide swath (IW) imaging mode and focused only on the data acquired in this mode. They present concepts are relevant for the less frequently used Extended Wide (EW) swath imaging mode imagery and for strip map mode imagery. The SLC images are used for automated processing and advanced exploitation, such as interferometric applications.

Digital Elevation Models: SRTM 30 m and 90 m. The Land Processes Distributed Active Archive Center (LP DAAC [3]) archives several types of high-quality terrain digital data, including Digital Elevation Models (DEM) data, which are used to describe the Earth's surface in terms of elevation. LP DAAC offers various DEM datasets such as the Shuttle Radar Topography Mission (SRTM) and ASTER Global DEM for free download on the EOSDIS data archive center's website or on Amazon S3. The proposed pipeline for DEM easy download use the SRTM 90 m Digital Elevation Database v4.1, elaborated by CGIAR-CSI [16], to download DEMs which have a resolution of 90 m at the equator and use the global datasets SRTM 30 m Global 1 arc second V003, elaborated by NASA and NGA, hosted on Amazon S3 to download DEMs which includes the global 1 arc second (30 m) number product. These DEMs play a crucial role in DinSAR pre-processing/geocoding (GTC) S-1 products.

3   Results and Discusions

This section discusses the proposed pipeline to generate a analysis ready datasets of DinSAR on Sentinel-1 Single Look Complex level-1 dual-polarized products that are suitable for deep learning. The results of the proposed pipeline applied to a specific area of interest are also presented.

3.1   Proposed Method: The SLCDeepETC Pipeline

In radar interferometry, the proposed pipeline (Eq. 1) consists of three stages (Fig. 1): Extract, Transform, and Cross-Platform Delivery Block, which process a given region of interest in-situ SLC level-1 dual-polarized (VV+VH) products as observations to generate Xarray tensors [10] of DinSAR time series ready data, including amplitude ($\tilde{A}$), coherence (coh) and phase (ϕ), followed by optional extended metadata. The pipeline allows bypassing the complexities of radar interferometry preprocessing and interpretation, instead focusing on training our model using ready data. Furthermore, the pipeline uses two extensively used open-source libraries, Mage and Ivy [11], providing the potential to orchestrate block activities and to create framework-agnostic DL libraries that can be used as a gateway between heterogeneous framework codebases for Deep Learning algorithms when training calls are made within a given runtime framework. The runtime framework codebases are defined as a platform (P). Each block in the pipeline is described as follows.

$$SLCDeepETC_{(OT)} : Observations \longmapsto Tensor[\tilde{A}, coh, \phi_{real}, \phi_{img}]_{VV+VH}^{P} \quad (1)$$

The Data Extraction Block: $Extract(E)$. The Extract block module is dedicated to ensuring the confidentiality, integrity, and availability of the data during their extraction from the provider. It streams raw Sentinel-1 SLC level-1 dual-polarized products with metadata for a specified geospatial area of interest from the Copernicus open-access hub using the embedded sentinelsat Python API. Additionally, the block automatically downloads the corresponding Digital Elevation Model (DEM) for the region of interest, which will be used for geometric [18] and radiometric [17] terrain correction. Subsequently, it stores these in a data lake with a dictionary file called "orderedDict" in a shared folder to represent the structure and content of the raw data. While there are several available API sources and open-source applications to extend the Extract block, it remains open for further extension by developers. As the output of the Extract Block, the user has a data lake of unstructured SLC files. To be utilized, these data require complex interferometry preprocessing to transform SLC level-1 dual-polarized products from a largely interpretive science to a quantitative tool. This has applications in cartography, geodesy, land cover characterization, and natural hazards, which is performed by the Transform block.

The Data Transformation Block: $Transform(T)$. The Transform block applies the synthetic aperture radar interferometry imaging technique [4] to unstructured data in the data lake from the Extract block, returning a time series of interferograms as a continuous flow of multidimensional Xarray tensor of amplitude, coherence, and phase radar signals. The interferograms are built following the given parametric period interval and apply adaptive method [19] to normalize all amplitude and split phase values to lie between 0 and 1, satisfying the deep learning model training recommendation. The Transform block outputs a stack of time series interferograms generated using the Diapotb Python

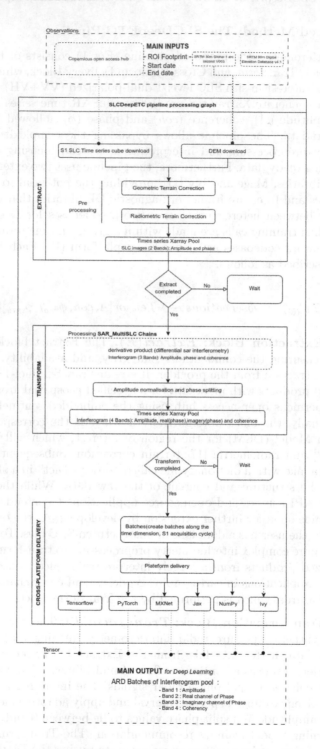

Fig. 1. A simplified schema illustrating the Single Look Complex Deep Learning ETC pipeline operational graph. All operational block remain accessible to the developer, allowing maximal control.

library and the Xarray-sentinel library as a backend, creating ready-to-use Xarray datasets as a data warehouse that lazily and efficiently map the data in terms of memory usage, disk/network access, with support for larger-than-memory and distributed data access via Dask or rioxarray/rasterio/GDAL. With the growth of deep learning frameworks on different code bases, having only ready data is not sufficient, and there is a need to deliver this data to whatever running platform is being used. This is the task of the Cross-platform Delivery block.

The Inter-framework Delivery Block: $Cross\_plateform\_delivery(C)$
The cross-platform delivery block links the Xarray tensor to DL libraries that support batch training/inference, using the open-source Python library xbatcher [9]. It provides a convenient Xarray accessor interface for iterating subsets of Xarray DataArrays/Datasets and PyTorch and TensorFlow Data Loaders for feeding tensor data into neural network models. The cross-platform delivery block also includes the open-source ivy-core framework, which supports popular libraries such as TensorFlow, PyTorch, MXNet, Jax, and NumPy, expanding the usability of other deep learning libraries for model training. The purpose of this block stage is to create a gateway between heterogeneous framework codebases for Deep Learning algorithms within the analysis-ready dataset. Providing an accessible platform, this approach considers the heterogeneity of deep learning frameworks and offers an innovative solution.

3.2 Experiments and Discusions

The section presents the experimentation of the SLCDeepETC pipeline for the on-demand generation of in-situ Analysis Ready Datasets for training deep learning models aimed at monitoring and predicting environmental phenomena in a region of interest.

Observed Area: To evaluate the effectiveness of the proposed SLCDeep-ETC pipeline, we conducted experiments on a selected area of interest - Mount Cameroon [1], situated on the littoral coast of the Central Africa zone, located at the Gulf of Guinea, as well as Gouache [21], situated in Bafoussam, the West Region of Cameroon, with a high occurrence of landslides and flooding in recent years. These natural phenomena have caused significant damage to the environment, infrastructure, and human lives in the region. Therefore, generating an Analysis Ready Dataset of interferometry derivative time series data on in-situ Single Look Complex Level-1 dual-polarized products for Deep Learning model training could be useful for monitoring and predicting these phenomena. The generation experiments were conducted on a computer with 8 GB of RAM, a 2.5 GHz Intel Core i5 processor, UMIOS operating system [24], and a 2-terabyte HDD with a 119 GB SSD storage capacity. The on-demand extraction period was one year, from 01-01-2018 to 31-12-2018, covering pre, during, and post-event periods.

On-Demand ARD Generated with SLCDeepETC: The experimental evaluations have highlighted several important aspects when comparing the SLCDeepETC pipeline with the SenSARP, snapista, pyroSAR, S1TILING, Diapason, and DiapOTB tools. In terms of computational complexity, it has been observed that the SLCDeepETC pipeline offers optimized performance, demonstrating faster processing times compared to the other evaluated tools. This improvement in efficiency is particularly notable when transforming raw data into analysis-ready datasets for deep learning and the ability to utilize data regardless of the deep learning algorithm framework used, highlighting the pipeline's capability to minimize the demand for computational resources while maintaining high levels of performance (Table 2 and Fig. 2).

Furthermore, the comparisons revealed that the SLCDeepETC pipeline presents more moderate memory requirements compared to some of the evaluated tools, suggesting better optimization of resource utilization while maintaining satisfactory performance. Additionally, the evaluation also highlighted the significant parallelization capacity of the SLCDeepETC pipeline, enabling effective handling of massive data transformation tasks (Fig. 3 and Fig. 4), a crucial factor in large-scale data processing scenarios.

Moreover, the simplified demonstration of how the SLCDeepETC pipeline optimizes computational complexity through the implementation of SAR data preprocessing algorithms to reduce spatial and temporal data size, integrated with adaptive filtering and data compression techniques, underscores the efficiency of this pipeline in reducing the demand for computational resources while preserving data quality (Fig. 5) and facilitating effective parallelization of processing tasks.

Thus, the experimental results confirm that the SLCDeepETC pipeline offers significant advantages in terms of computational complexity compared to SenSARP, snapista, pyroSAR, S1TILING, Diapason, and DiapOTB. These observations emphasize the relevance and efficiency of the SLCDeepETC pipeline as a potential leading solution for generating analysis-ready datasets for deep learning, with direct implications for its application in complex data processing environments. Overall, the results obtained under the specified hardware setup support in this study highlight the SLCDeepETC pipeline potential and the feasibility of the approach, making it applicable to a wider range of similar hardware configurations.

Limits and Further Research. It should be noted that the presented method also has limits. The user is responsible for providing crucial parameters to the applications, specifically the pre-processing, DInSAR, and post-processing fields of the DiapOTB chain. In future development, those parameters could be extracted from the metadata. In the same way, the pipeline for interferometric processing is computed within the DiapOTB pipeline, but does not takes in

Table 2. Description of the generated ARD datasets

id	Roi	Product type	Product number	Dataset size	Batches number	DEM number	times series	Inteferogram and coherance data number
Output	mount cameroon	SLC	92	645,57G	46	3	20180101–20181231	46
	gouache	SLC	30	215,14G	15	3	20180101–20181231	15
comments			The product is available on the Copernicus Hub, with an acquisition cycle of six days for Mount Cameroon and twelve days for Gouache		We created batches based on the time dimension, with an interval of six days for Mount Cameroon, and twenty-four days for Gouache, respectively	One per subband (IW1, IW2, IW3) always this number	We acquired data for one year, with two image acquisitions every six days for Mount Cameroon, and every twenty-four days for Gouache	Interferograms and coherence data were obtained for Mount Cameroon and Gouache, with a cycle of six and twenty-four days between each acquisition, respectively

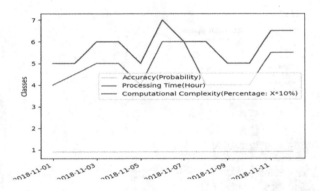

Fig. 2. Processing Performance Evaluation by Product Acquisition Date: This graph shows the performance when processing the pipeline in our experimental environment according to the evaluation criteria shown in Table 1.

account the memory consumed internally by DiapOTB python chains during the execution of the computational graph: the user is thus responsible of fine tuning using the DiapOTB application engine parameters.

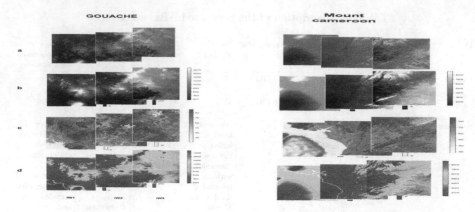

Fig. 3. Digital Elevation Models (DEMs) of the region of interest (ROI) are downloaded by the pipeline per sub-band (IW1, IW2, IW3). The left image shows the DEM of Gouache, while the right image shows the DEM of Mount Cameroon: (a) DEM with Contour Visualization, (b) DEM visualization with RichDEM Package [2], (c) Slope Visualization from DEM [8], (d)Depression filled DEM [26].

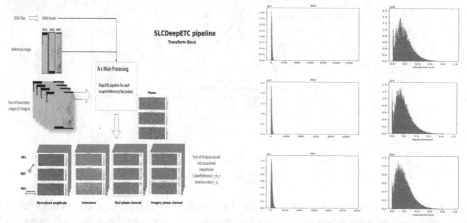

Fig. 4. The SLCDeepETC Transform (T) Block workflow embeds the Python DiapOtb tools as a set of synthetic aperture radar interferometry imaging techniques to obtain amplitude, coherence, and phase over time, capturing temporal changes in the observed scenes (Gouache) for various applications.

Fig. 5. Amplitude data disparity: left raw (A) before pipeline processing and right ($\tilde{A}$) after pipeline processing, per sub-band IW1, IW2 and IW3 on Gouache.

4   Conclusion

The SLCDeepETC Pipeline proposed in this study offers potential for generating on-demand Analysis Ready Data of DInSAR products suitable for deep learning applications. The pipeline incorporates three operational stages and generates an on-demand time series tensor for a chosen observed area of interest using extensively used open-source libraries. The pipeline provides flexibility for deriving secondary data from primary data and accommodating the heterogeneity of DL frameworks. The SLCDeepETC pipeline architecture can serve as a pioneer in establishing a multiplatform public Analysis Ready Dataset of DInSAR products for SLC-type product DL training analysis, enabling researchers to perform advanced DL analyses that were previously impossible. The pipeline remains open for further extension, making it a valuable resource for the scientific community.

Acknowledgments. All satellite data were provided by the European Space Agency (ESA) and are publicly available online through the Copernicus Open Access Hub at https://scihub.copernicus.eu/. The DEM data were provided by the Land Processes Distributed Active Archive Center (LP DAAC) and are publicly available via the global datasets SRTM 30 m Global 1 arc second V003, elaborated by NASA and NGA, and hosted on Amazon S3, as well as the SRTM 90 m Digital Elevation Database v4.1, elaborated by CGIAR-CSI.

Disclosure of Interests. No potential conflict of interest was reported by the authors.

References

1. Annet, E.: Le mont cameroun. Revue d'Ecologie, Terre et Vie **10**, 611–621 (1931)
2. Barnes, R.: RichDEM: terrain analysis software (2016). http://github.com/r-barnes/richdem
3. Behnke, J., Doescher, C.: Land processes distributed active archive center (LP DAAC) 25th anniversary recognition "a model for government partnerships". LP DAAC "history and a look forward". In: LP DAAC Recognition Meeting. No. GSFC-E-DAA-TN26383 (2015)
4. Bürgmann, R., Rosen, P.A., Fielding, E.J.: Synthetic aperture radar interferometry to measure earth's surface topography and its deformation. Annu. Rev. Earth Planet. Sci. **28**(1), 169–209 (2000)
5. Durand, P., Pourthie, N., Usseglio, G., Tison, C.: DiapOTB: a new open source tool for differential SAR interferometry. In: EUSAR 2021; 13th European Conference on Synthetic Aperture Radar, pp. 1–4. VDE (2021)
6. Friedt, J.M., Abbé, P.: Parler à un radar spatioporté: traitement et analyse des données de sentinel-1. GNU/Linux Mag. **246**, 18 (2021)
7. Hong, D., et al.: More diverse means better: multimodal deep learning meets remote-sensing imagery classification. IEEE Trans. Geosci. Remote Sens. **59**(5), 4340–4354 (2020)
8. Horn, B.K.: Hill shading and the reflectance map. Proc. IEEE **69**(1), 14–47 (1981)
9. Jones, M., Hamman, J.J., Leong, W.J.: Xbatcher - a Python package that simplifies feeding Xarray data objects to machine learning libraries. In: 103rd AMS Annual Meeting. AMS (2023)
10. Kolda, T.G., Bader, B.W.: Tensor decompositions and applications. SIAM Rev. **51**(3), 455–500 (2009)

11. Lenton, D., Pardo, F., Falck, F., James, S., Clark, R.: Ivy: templated deep learning for inter-framework portability. arXiv preprint arXiv:2102.02886 (2021)
12. Ma, L., Liu, Y., Zhang, X., Ye, Y., Yin, G., Johnson, B.A.: Deep learning in remote sensing applications: a meta-analysis and review. ISPRS J. Photogramm. Remote. Sens. **152**, 166–177 (2019)
13. Massonnet, D., Adragna, F.: Description of the Diapason software developed by CNES current and future applications. In: ERS SAR Interferometry, vol. 406, p. 202 (1997)
14. Minh, D.H.T., et al.: Deep recurrent neural networks for winter vegetation quality mapping via multitemporal SAR Sentinel-1. IEEE Geosci. Remote Sens. Lett. **15**(3), 464–468 (2018)
15. Otamendi, U., Azpiroz, I., Quartulli, M., Olaizola, I.: Integrating pre-processing pipelines in ODC based framework. In: IGARSS 2022-2022 IEEE International Geoscience and Remote Sensing Symposium, pp. 4094–4097. IEEE (2022)
16. Reuter, H.I., Nelson, A., Jarvis, A.: An evaluation of void-filling interpolation methods for SRTM data. Int. J. Geogr. Inf. Sci. **21**(9), 983–1008 (2007)
17. Small, D.: Flattening gamma: radiometric terrain correction for SAR imagery. IEEE Trans. Geosci. Remote Sens. **49**(8), 3081–3093 (2011)
18. Small, D., Schubert, A.: Guide to Sentinel-1 geocoding. Remote Sensing Lab. University of Zurich (RSL), Zürich, Switzerland, Technical report. UZHS1-GC-AD (2019)
19. Sun, X., Zimmer, A., Mukherjee, S., Kottayil, N.K., Ghuman, P., Cheng, I.: DeepInSAR-a deep learning framework for SAR interferometric phase restoration and coherence estimation. Remote Sens. **12**(14), 2340 (2020)
20. Truckenbrodt, J., Cremer, F., Baris, I., Eberle, J., et al.: PyroSAR: a framework for large-scale SAR satellite data processing. In: Proceedings of the Big Data from Space, Munich, Germany, pp. 19–20 (2019)
21. Tsoata, F.T., Yemmafouo, A., Ngouanet, C.: Cartographie de la susceptibilité aux glissements de terrain à bafoussam (cameroun). approche par analyse multicritère hiérarchique et système d'information géographique. Revue internationale de géomatique, aménagement et gestion des ressources (2020)
22. Wang, L., Scott, K.A., Xu, L., Clausi, D.A.: Sea ice concentration estimation during melt from dual-pol SAR scenes using deep convolutional neural networks: a case study. IEEE Trans. Geosci. Remote Sens. **54**(8), 4524–4533 (2016)
23. Weiß, T., Fincke, T.: SenSARP: a pipeline to pre-process Sentinel-1 SLC data by using ESA SNAP Sentinel-1 toolbox. J. Open Source Softw. **7**(69), 3337 (2022)
24. Welaab: UMI-OS: a reliable and stable operating system for an ergonomic user experience (2019). https://weloobe.com/products/umios
25. Xing, H., Meng, Y., Wang, Z., Fan, K., Hou, D.: Exploring geo-tagged photos for land cover validation with deep learning. ISPRS J. Photogramm. Remote. Sens. **141**, 237–251 (2018)
26. Zhou, G., Sun, Z., Fu, S.: An efficient variant of the priority-flood algorithm for filling depressions in raster digital elevation models. Comput. Geosci. **90**, 87–96 (2016)
27. Zhu, M., He, Y., He, Q.: A review of researches on deep learning in remote sensing application. Int. J. Geosci. **10**(1), 1–11 (2019)

Robustness of Image Classification on Imbalanced Datasets Using Capsules Networks

Steve Onana[1]([✉]), Diane Tchuani[2], Claude Tinku[1], Louis Fippo[1], and Georges Edouard Kouamou[1]

[1] University of Yaounde I, Yaoundé, Cameroon
onana.stevealain@gmail.com , louis.fippo@univ-yaounde1.cm
[2] University of Ebolowa, Ebolowa, Cameroon

Abstract. The purpose of this study is to assess the robustness of the capsule networks for the image classification tasks on imbalanced dataset. In fact, in many real-life situations, the distribution of samples is skewed, with representatives of certain classes appearing much more frequently. This raises a problem for learning algorithms, as they will be biased in favor of the majority group. At the same time, the minority class is usually the most important, because, despite its rarity, it may carry important and useful information. In this work, we firstly design a capsule networks architecture that we apply to a medical dataset extremely imbalanced provided by the Broad Institute, then we randomly subsampled that dataset and we compare the result to CNN architectures. To gauge our model's ability to perform well on diverse datasets, we tested it on a specifically configured version of the CIFAR10 dataset. This version has a class imbalance ratio of 0.1, meaning some classes have significantly fewer data points than others. This mimics the real-world scenario of imbalanced datasets. We then compared our model's performance on this dataset with the findings of recent research in the field. To evaluate results, we use the following metrics: Accuracy, Precision, Recall, and F1-score. The results of experiments show that capsules networks can give good results on highly imbalanced datasets (BBBC41 and BBBC042) and with small observations and can be in certain instance better than convolutional neural networks. While our Capsule network architecture achieved a recall of 0.7 on the imbalanced CIFAR10 dataset, recent studies utilizing focal loss and complement cross-entropy with convolutional neural networks (CNNs) have demonstrated superior performance. These findings highlight the impact of both model architecture and data characteristics. Our model achieved better results on datasets with more severe imbalances (1:700 and 1:15) compared to CIFAR10's milder imbalance (1:10). This underscores the importance of considering both model and dataset factors when evaluating performance on imbalanced data. In conclusion, Capsule networks show promise for imbalanced learning, but they are not a universal solution. Further research is needed to explore their effectiveness across a wider range of imbalanced data scenario.

P. Melatagia Yonta et al. (Eds.): CRI 2023, CCIS 2085, pp. 53–68, 2024.
https://doi.org/10.1007/978-3-031-63110-8_5

Keywords: Imbalanced datasets · Image classification · Capsule
Networks · Convolution neural Networks

1   Introduction

Image classification, a cornerstone of machine learning and deep learning, plays
a critical role in various applications. It involves assigning a category label to
an image based on its content. However, traditional classification algorithms
assume a balanced dataset, where all classes are represented equally. In reality,
real-world data often exhibits class imbalance, meaning some classes appear
significantly more frequently than others. This imbalance poses a challenge for
learning algorithms, as they become biased towards the majority class, neglecting
the potentially crucial information held by the underrepresented (minority) class.
This bias is particularly problematic for online learning models trained on live
streams, where the majority class is constantly encountered. To address this
issue, researchers have developed various techniques for effective classification
on imbalanced datasets [2].

There are methods based on data. These methods apply changes to the
dataset. Among these methods we can mention, Data augmentation(applications
of matrix transformations to the training data) [10]. Sampling methods, there
are several sampling methods, random oversampling, Smote, Adasyn [15,17],
random subsampling, Tomek Link [16]. Some studies also use GAN as oversampling [19] techniques often when the dataset is composed of images. In addition
to these methods, which have the disadvantage of performing transformations
on the dataset, There are also algorithm methods. These methods can be used
by creating robust learning algorithms. These include, Ensemble Learning [6,7],
Transfer Learning [8], Weighted Loss function (focal loss [4], Maximum margin
Loss [5]), Complement Cross Entropy Loss [19]. It's also possible to have hybrid
methods which are the combination of all these methods. All These methods
achieve great results and improve the overall performance when they were tested.

The year 2017 saw the introduction of a novel architecture: capsule networks
[1]. This approach demonstrated impressive performance on image classification tasks, particularly with the balanced MNIST dataset. Capsule networks
offer several advantages over traditional convolutional neural networks (CNNs).
Notably, they achieve high performance with fewer parameters and possess the
unique ability to preserve feature relationships through the routing agreement
algorithm, replacing the pooling layer. Additionally, the margin loss function
employed further strengthens this architecture. These characteristics motivated
us to investigate the robustness of capsule networks when applied to imbalanced
datasets. This paper examines the effectiveness of capsule networks on imbalanced datasets. While we demonstrate that capsule networks, combined with a
simple data augmentation technique, can achieve significant results in certain
cases, we emphasize that they are not a universal solution for all imbalanced
data problems. We will evaluate the performance of capsule networks on two
imbalanced datasets and compare our findings with recent research on imbalanced data management strategies. The remainder of this paper is structured as

follows: Sect. 2 describe the methods used in this paper and present the datasets. In Sect. 3, we discuss and present the experimental settings, the environment, the experiments, and the results for this paper. Results are discussed in Sect. 3.3.

2   Methods

In this section, we describe convolutional neural networks and capsules networks architectures. We describe the datasets used.

2.1   Convolutional Neural Networks

Convolution neural networks are an architecture created by [9]. That architecture was mainly designed for image classification and tried to mimic how the visual cortex works. Convolutional neural networks since 2012 with the presence of big data and power computing has became the state of the art for image classification tasks. They are made of three layers.

- Convolution Layer. That layer with the pooling layer are responsible for feature extraction. Filters apply cross-correlation product to an image in order to detect important features then apply the activation function and pass the result to the pooling layer.
- Pooling Layer. That Layer is used to downsample the output of the convolution layer. We can use average or max functions to perform pooling. Convolution and pooling can be performed several times if the model needs to have more abstract features.
- Fully Connected Layer. After all the convolution and pooling tasks are performed the final output is sent to the fully connected layers for classification sake.

While Convolutional Neural Networks (CNNs) have dominated image classification tasks, their pooling layers present a limitation. Specifically, pooling can discard spatial relationships between features, hindering the networks' ability to achieve equivariance and invariance to rotations. To address this limitation, Capsule Networks were introduced. In this study, we compare these two approaches. We utilize three CNN architectures: a baseline model, and two pre-trained models, DenseNet121 and InceptionV3. These pre-trained models achieved success in image classification on the ImageNet dataset. By employing these architectures, we aim to comprehensively evaluate Capsule Networks against CNNs on a medical dataset.

2.2   Capsule Networks

A capsule networks is a type of artificial neural networks created by [1] to address several limitations of convolutional neural networks (maintain hierarchical relations). A capsule according to [1] is a group of neurons that can be represented as an abstract vector. That vector represents the instantiation parameters (hue, width, height, texture, velocity...) of a specific type of entity such as an object or an object part. The length(norm) of that vector represents the probability that the entity exists and its orientation represents the instantiation parameters. So capsule networks is made of convolution layer to extract features and capsules layers(composed of capsules). There must be at least two capsules layers(two levels). Active capsules at one level make predictions, via transformation matrices 1, for the instantiation parameters of higher-level capsules. And the active layer of the higher level capsules are chosen by the routing algorithm.

$$v_j = squash(\sum_i c_{ij} w_{ij} * u_i) \tag{1}$$

The squash function is a mathematical function or a nonlinear activation function that takes as input a vector and outputs the norm of that vector bounded between 0 and 1. This norm is the probability that the entity exists.
v_j: is the output of the capsule j
w_{ij}: is the weight matrix between the capsule i and the capsule j. The values of the elements of this matrix are obtained by backpropagation.
u_i: is the output of the capsule j c_{ij}: are coupling coefficients that are determined by the iterative dynamic routing process. They represent the strength of the coupling between two capsules.

2.3   Dynamic Routing Algorithm

The fact that capsule networks can preserve the hierarchy between part and entity is due to the routing agreement algorithm. It's an iterative algorithm that takes as arguments the number of iterations. The capsule level and the prediction vector. As shown in the following figure (Fig. 1).

Procedure 1 Routing algorithm.

1: **procedure** ROUTING($\hat{u}_{j|i}, r, l$)
2:     for all capsule i in layer l and capsule j in layer $(l+1)$: $b_{ij} \leftarrow 0$.
3:     **for** r iterations **do**
4:         for all capsule i in layer l: $c_i \leftarrow \text{softmax}(b_i)$ ▷ softmax computes Eq. 3
5:         for all capsule j in layer $(l+1)$: $s_j \leftarrow \sum_i c_{ij} \hat{u}_{j|i}$
6:         for all capsule j in layer $(l+1)$: $v_j \leftarrow \text{squash}(s_j)$ ▷ squash computes Eq. 1
7:         for all capsule i in layer l and capsule j in layer $(l+1)$: $b_{ij} \leftarrow b_{ij} + \hat{u}_{j|i}.v_j$
    return v_j

Fig. 1. routing Agreement algorithm [1]

2.4   Architecture of Capsule Networks and Loss Function

Capsules networks architecture can be divided into two big parts:

– Encoder. The main role of the Encoder is the classification and is made of:

- Convolution layers that are used for feature extraction. We can use 2 or more convolution layers. In this study we use five convolution layers inspired by the work of [3]. The more we have convolution layers the more the features are diverse.
- Capsules layers. We just use two capsule layers for this study. A primary capsules layers which will take the output after the final convolution layer and transform the neurons in capsules. The second layer of capsules is used for classification. The output of that layer will be obtained after the routing algorithm.
- The decoder is composed of fully connected neural networks layers. The decoder has two roles. The first role is the reconstruction role that role gives the autoencoder functionality to capsule networks. the second role is the regularization function that gives the architecture the possibility to avoid overfitting. For this study the reconstruction wasn't relevant we get the best result without the decoder part (Figs. 2 and 3).

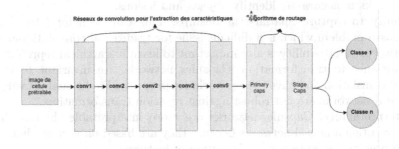

Fig. 2. Encoder architecture

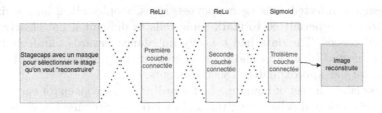

Fig. 3. Decoder Architecture [1]

The loss function of the capsule layers depends on the facts that we use the decoder or not in our model.

The general formula of the total loss

$$L = \sum_k L_k = T_k max(0, m^+ - \|V_c\|)^2 + \lambda(1 - T_k)max(0, \|V_c\| - m^-)^2 + \alpha.mse(T_k, T_{rk})$$

(2)

V_c: output of the predicted capsule
T_k: label
m^+: upper bound
m^-: lower bound
T_{rk}: label's reconstruction
α: reconstruction coefficient

This study on the assessment of capsules networks on imbalanced datasets is motivated by the fact that capsules networks have the following properties.

- Robustness to Noise and occlusions: Medical images often contain noise and occlusions, which can make it difficult for traditional convolutional neural networks to accurately identify objects and lesions.
- Ability to capture Spatial relationships: This can be helpful to solve the Picasso problem which is a difficult issue for traditional convolutional neural networks. That Ability is also important to learn hierarchical representations
- Invariance to transformation: Capsules networks are invariant to rotation and translation and have the equivariance property. Traditional convolutional neural networks are not robust against rotation transformation.
- Interpretability: Capsule networks are more interpretable than traditional convolutional neural networks because they are based on vectors that encode the presence, orientation, and position of features.

2.5   Datasets

The first dataset is a medical dataset that was created for a study by [11]. The dataset consists of two classes of uninfected cells (RBCs and leukocytes) and four classes of infected cells (gametocytes, rings, trophozoites, and schizonts). Annotators were permitted to mark some cells as difficult if not clearly in one of the cell classes. The data had a heavy imbalance towards uninfected RBCs versus uninfected leukocytes and infected cells, making up over 95% of all cells. This dataset has approximately 73000 elements.

A class label and set of bounding box coordinates were given for each cell. For all data sets, infected cells were given a class label by Stefanie Lopes, a malaria researcher at the Dr. Heitor Vieira Dourado Tropical Medicine Foundation hospital, indicating the stage of development or marked as difficult.

The segmentation process is not relevant to this study. For this study we will call the initial dataset BBBC041, that dataset has an imbalance ratio of 1:700. We will derive a new dataset from the initial dataset by applying the random subsampling (only on the majority class) in order to reduce the imbalance ratio. We will call the subsampled dataset BBBC042. The dataset BBBC042 will have 3800 elements with an imbalanced ratio of 1:15. We obtain the BBBC042 dataset by removing randomly images of the majority class until we get our desired number of observations.

In order to see if our model can work on some type of balanced medical networks and validate our intuition that capsules networks performs well on medical either the dataset is balanced or imbalanced we use a balanced dataset for malaria that dataset has 2 classes.

The distribution of the two datasets is shown in the following figures (Figs. 4 and 5):

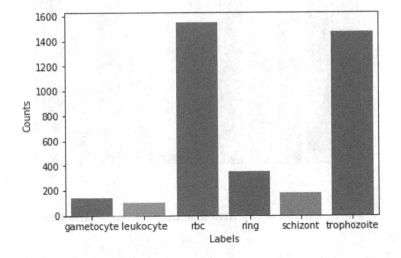

Fig. 4. Box plot of the dataset BBBC042 with an imbalance ratio 1:15

To benchmark our architecture against recent approaches, we will employ a second dataset: the CIFAR10 long-tail dataset with a 1:10 class imbalance ratio [20]. This dataset consists of ten classes, with each observation being a $32 \times 32 \times 3$ image. Additionally, we will leverage the findings from the study [21] on balanced CIFAR10 utilized with capsule networks (Fig. 6).

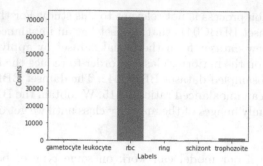

Fig. 5. Box plot of the dataset BBBC041 with an imbalance ratio 1:700

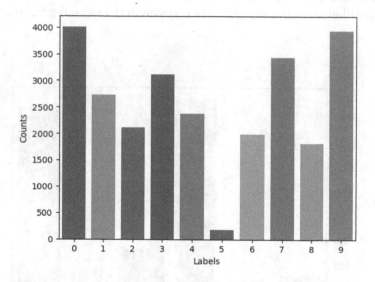

Fig. 6. Bar plot of CIFAR10 long trail imbalance ratio 1:10

We evaluated our capsule network's performance on a balanced medical dataset from NIH [22] to validate our hypothesis that it generalizes well to both balanced and imbalanced medical data. Specifically, we employed a balanced malaria dataset with two classes for this initial assessment (Fig. 7).

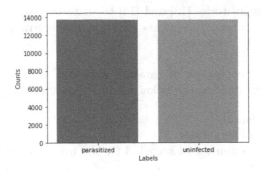

Fig. 7. Bar plot of NIH dataset

3   Experiments and Results

This section first briefly overviews the experimental setup and implementation details and then presents experimental results per each imbalanced image dataset.

3.1   Training Process

- We load the datasets from the disk.
- We divide the initial and the subsampled datasets into three parts training/validation/test. To avoid recollection (overfitting) of the model thus we will have a good evaluation of the model.
- We train our models without with data augmentation. The data augmentation methods used are random flip and random rotation
- We use Capsule networks with and without the decoder on the first dataset and on the subsampled dataset.
- We use pre-trained convolutional neural networks (transfer learning) on the initial dataset and on the subsampled dataset.

Because we use an imbalanced dataset Accuracy used alone as a metric is not relevant because it's a grouping metric. The majority class will dictate the value of the Accuracy. So we will have to use recall, precision, and F1-score to have a good evaluation of our model. In order to compare with other works for the second dataset the only metric used is the recall.

3.2   Experiments

For the experiments, we use the framework TensorFlow 2.11 as a backend of Keras and the programming language Python3. We take inspiration from the following GitHub repositories [13] and [14] to write our codes. We also use the examples available on the TensorFlow website. For hardware, we use Google Colab and a personal computer.

Our first model composed of capsules networks applied to the datasets BBBC041 and BBBC042 has the following hyperparameters:

- Image size: (96, 96, 3).
- Convolution layers kernel size: 5
- Convolution layers number of kernel: 3 for the first convolution layer and 64 for the remainder
- padding is valid
- stride: 1 for the first convolution layer 2 for the remainder
- dropout parameter is 0.4
- batch normalization layer
- The optimizer used is ADAM
- The Learning rate 0.0001
- The batch size is 64
- The number of epochs is 50.
- The number of iterations of the dynamic routing algorithm is 4.

Our second model composed of capsules networks applied to the CIFAR10 long trail dataset has the following hyperparameters:

- Image size: (96, 96, 3).
- Convolution layers kernel size: 5
- Convolution layers number of kernel: 3 for the first convolution layer and 64 for the remainder
- padding is valid
- stride: 1 for the first convolution layer 2 for the remainder
- dropout parameter is 0.4
- batch normalization layer
- The optimizer used is ADAM
- The Learning rate 0.001
- The batch size is 256
- The number of epochs is 300.
- The number of iterations of the dynamic routing algorithm is 4.

These hyperparameters have been selected based on the trade-off between the performances, the computing power, and the learning time.

In order to see how our model perform we generate confusion matrix from which we can derive our metrics. Confusion matrix results come from the test set (Figs. 8 and 9).

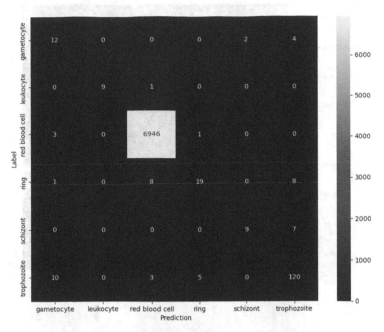

(a) Confusion Matrix of the dataset BBBC041

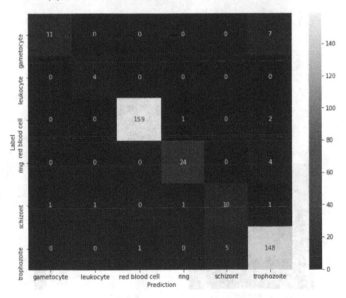

(b) Confusion matrix of subsampled dataset BBBC042

Fig. 8. Confusion Matrix for Our medical dataset and it's subsampled version.

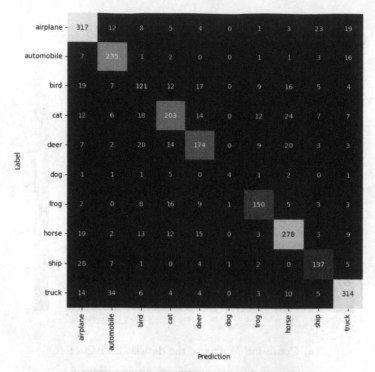

(a) Confusion Matrix of CIFAR10 long trail dataset

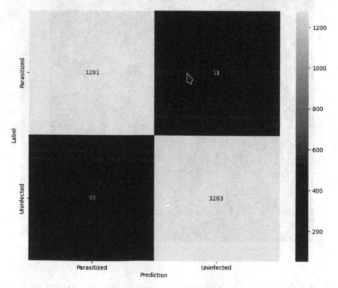

(b) Confusion Matrix of NIH dataset

Fig. 9. (a) Confusion Matrix of CIFAR10 long trail dataset (b) Confusion Matrix of NIH dataset

3.3 Discussions

Our experiments demonstrated the effectiveness of capsule networks for handling highly imbalanced datasets (Table 1, line 3). Capsule networks consistently outperformed convolutional networks, regardless of decoder usage. This was true for both the original BBBC041 dataset and the subsampled BBBC042 dataset (Tables 1 and 2, lines 3 & 4 and lines 2 & 3). Interestingly, capsule networks achieved better performance on the smaller dataset, sacrificing some accuracy

Table 1. Initial dataset: BBBC041. The naming convention for capsXX. The first number represents the usage of the decoder if 1 we use that augmentation if 0 we don't use the decoder and the second number represents the usage of the data augmentation.

Dataset	Model	Decoder	Augmentation	Accuracy	Precision	Recall	F1score
BBBC041	DENSENET	-	yes	0.98	0.36	0.34	0.24
BBBC041	INCEPTION	-	yes	0.99	0.41	0.39	0.39
BBBC041	**Caps01**	**no**	**yes**	**0.99**	**0.82**	**0.75**	**0.77**
BBBC041	Caps11	yes	yes	0.99	0.78	0.77	0.77
BBBC041	Caps00	no	no	0.99	0.71	0.60	0.63
BBBC041	CNN	–	yes	0.99	0.72	0.66	0.68

Table 2. Subsampled dataset: BBBC042

Dataset	Model	Decoder	Augmentation	Accuracy	Precision	Recall	F1score
BBBC042	INCEPTION	-	yes	0.86	0.66	0.47	0.48
BBBC042	Caps11	yes	yes	0.83	0.69	0.64	0.65
BBBC042	**Caps01**	**no**	**yes**	**0.94**	**0.87**	**0.85**	**0.85**

Table 3. This table compare the result obtained by our architecture to the results obtained with complement cross entropy loss and focal loss. We can see that our architecture doesn't perform better than those architecture with that dataset

Dataset	Model	Decoder	Augmentation	Recall
CIFAR10-LT	**Caps01**	**no**	**yes**	**0.7**
CIFAR10-LT	Resnet34+CCE [18]	–	–	0.8837
CIFAR10-LT	Focal Loss [18]	–	–	0.8616

Table 4. This table show us how capsule networks behave on balanced dataset. The medical dataset NIH achieves great results meanwhile balanced CIFAR10 is under 0.7.

Dataset	Model	Decoder	Augmentation	Recall
NIH	Caps01	no	yes	0.95
CIFAR10	CapsNet [18]	yes	yes	0.68

but improving other metrics. This suggests their suitability for limited data scenarios.

Data augmentation's importance is evident when comparing lines 4 and 5 of Table 1. Without augmentation, performance suffered. However, our capsule architecture applied to the CIFAR10 long tail dataset yielded results surpassed by other architectures (Table 3). While the CIFAR10 results are not insignificant from a broader perspective, they fall short of recent benchmarks for this dataset. Our findings, as shown in Table 4, suggest that Capsule Networks may be more suited for specific datasets. While performance on the balanced CIFAR10 dataset was modest, they achieved excellent results on the NIH medical dataset. This indicates that Capsule Networks might not generalize well to all balanced datasets but have the potential to excel in medical image analysis. In summary, our study demonstrates that Capsule Networks can be effective for both balanced datasets (in the case of the NIH data) and imbalanced medical datasets (BBBC041 and BBBC042).

4   Conclusion

This study investigated the robustness of capsules networks for image classification on imbalanced datasets. We compared capsules networks and convolutional neural network (CNN) architectures on two datasets, applying established techniques to address imbalanced data bias.

The results were promising. Capsules networks outperformed convolutional neural networks on highly imbalanced medical datasets. However, our capsules network architecture did not surpass recent results on the CIFAR10 long-tail dataset. This suggests that Capsules networks are not a universal solution, and their performance depends on the dataset characteristics. We achieved better results with lower imbalance ratios on datasets BBBC041 and BBBC042, indicating potential in this area. We also found that even with balanced datasets capsules networks doesn't always get greats results and that all medical datasets used in this study give promising results whether they were balanced or not. Encouraged by these findings, we propose further research directions:

- Dataset suitability and explainability: Investigate if specific data features or descriptors favor Capsules networks.
- Loss functions: Implement the recently proposed complement cross-entropy loss function for capsules networks or design a new loss function for capsules networks
- Network architecture: Experiment with deeper Capsules Networks architectures (more than two capsules layers) and alternative data augmentation techniques beyond random flipping and rotation.

References

1. Sabour, S., Frosst, N., Hinton, G.E.: Dynamic routing between capsules (2017)
2. Krawczyk, B.: Learning from imbalanced data: open challenges and future directions. Prog. Artif. Intell. **5**, 221–232 (2016)
3. Madhu, G., Govardhan, A., Srinivas, B.S., Patel, S.A., Rohit, B., Bharadwaj, B.L.: Capsule networks for malaria parasite classification: an application oriented model. In: 2020 IEEE International Conference for Innovation in Technology (INOCON) (2020)
4. Lin, T.-Y., Goyal, P., Girshick, R., He, K., Dollár, P.: Focal loss for dense object detection. IEEE Trans. Pattern Anal. Mach. Intell. **42**(2), 318–327 (2020)
5. Kang, H., Vu, T., Yoo, C.D.: Learning imbalanced datasets with maximum margin loss. In: 2021 IEEE International Conference on Image Processing (ICIP) (2021)
6. Liu, Z., et al.: Self-paced ensemble for highly imbalanced massive data classification. In: 2020 IEEE 36th International Conference on Data Engineering (ICDE) (2020)
7. Gao, Y., Wang, M., Zhang, G., Zhou, L., Luo, J., Liu, L.: Cluster-based ensemble learning model for aortic dissection screening. Int. J. Environ. Res. Public Health **19**, 5657 (2022)
8. Liu, Y., et al.: Imbalanced data classification: using transfer learning and active sampling. Eng. Appl. Artif. Intell. **117**, 105621 (2023)
9. Bengio, Y., Lecun, Y.: Convolutional networks for images, speech, and time-series (1995)
10. Shorten, C., Khoshgoftaar, T.M.: A survey on image data augmentation for deep learning. J. Big Data **6**, 60 (2019)
11. Ljosa, V., Sokolnicki, K., Carpenter, A.: Annotated high-throughput microscopy image sets for validation. Nat. Methods **9**, 637 (2012)
12. Zhuang, F., et al.: A comprehensive survey on transfer learning. Proc. IEEE **109**(1), 43–76 (2021)
13. Guo, X.: A Keras implementation of CapsNet in NIPS2017 paper "Dynamic Routing Between Capsules", Github repository. https://github.com/XifengGuo/CapsNet-Keras
14. Do, D.: Diseases Detection from NIH Chest X-ray data. Github repository. https://github.com/DoDuy/Lung-Diseases-Classifier
15. Chawla, N.V., Bowyer, K.W., Hall, L.O., Kegelmeyer, W.P.: SMOTE: synthetic minority over-sampling technique (2011)
16. Kulkarni, A., Chong, D., Batarseh, F.A.: Foundations of data imbalance and solutions for a data democracy (2020)
17. He, H., Bai, Y., Garcia, E., Li, S.: ADASYN: adaptive synthetic sampling approach for imbalanced learning. In: Proceedings of the International Joint Conference on Neural Networks (2008)
18. Kim, Y., Lee, Y., Jeon, M.: Imbalanced image classification with complement cross entropy. Pattern Recogn. Lett. **151**, 33–40 (2021)
19. Belderbos, I., de Jong, T., Popa, M.: GANs based conditional aerial images generation for imbalanced learning. In: Pattern Recognition and Artificial Intelligence: Third International Conference, ICPRAI 2022, Paris, France, 1–3 June 2022, Proceedings, Part II (2022)

20. Krizhevsky, A., Nair, V., Hinton, G.: CIFAR-10 (Canadian Institute for Advanced Research). http://www.cs.toronto.edu/~kriz/cifar.html
21. Nair, P., Doshi, R., Keselj, S.: Pushing the limits of capsule networks (2021)
22. Ragb, H.K., Dover, I.T., Ali, R.: Deep convolutional neural network ensemble for improved malaria parasite detection (2020)

Application of the Multilingual Acoustic Representation Model XLSR-53 for the Transcription of Ewondo

Yannick Yomie Nzeuhang[1(✉)], Paulin Melatagia Yonta[1,2], and Benjamin Lecouteux[3]

[1] Department of Computer Sciences, University of Yaounde I, Yaoundé, Cameroon
yynzeuhang@gmail.com
[2] IRD, UMMISCO, 93143 Bondy, France
[3] Univ. Grenoble Alpes, CNRS, Grenoble INP, LIG, 38000 Grenoble, France

Abstract. Recently popularized self-supervised models appear as a solution to the problem of low data availability via parsimonious learning transfer. We investigate the effectiveness of these multilingual acoustic models, in this case wav2vec 2.0 XLSR-53, for the transcription task of the Ewondo language (spoken in Cameroon). The experiments were conducted on 24 min of speech constructed from 103 read sentences. Despite a strong generalization capacity of multilingual acoustic model, preliminary results show that the distance between XLSR-53 embedded languages (English, French, Spanish, German, Mandarin, . . .) and Ewondo strongly impacts the performance of the transcription model. The highest performances obtained are around 70.8% on the WER and 28% on the CER. An analysis of these preliminary results is carried out and then interpreted; in order to ultimately propose effective ways of improvement.

Keywords: Low resource language · Self-supervised model · XLSR-53 · Transcription · Ewondo

1 Introduction

Self-supervised learning is a deep learning method for learning robust representations from unlabeled data. The main idea is to automatically generate labels for a simple pretext task, enabling the model to better understand the given structure, and then to use this learned information for a more complex target task. This method has recently been widely illustrated in speech processing, notably by the multilingual acoustic model **wav2vec 2.0 XLSR-53** [4], which delivers impressive results for automatic speech recognition (ASR) tasks, even on small datasets. By these fact this model presents itself as a solution for low resources languages for which automatic speech processing tasks are difficult to address by deep learning, due to the difficulty of building a large dataset. Ewondo, language from central Cameroon falls into this category of language.

Our aim is to evaluate effectiveness of multilingual acoustic model on Ewondo, which has the particularity of being tonal. To achieve this goal, we have

P. Melatagia Yonta et al. (Eds.): CRI 2023, CCIS 2085, pp. 69–79, 2024.
https://doi.org/10.1007/978-3-031-63110-8_6

built several ASR models based on *wav2vec 2.0* [2] in various configurations, we have evaluated the performance on word error rate (WER) and character error rate (CER). Our contribution in this paper is twofold: 1) The construction of a basic speech recognition model for Ewondo 2) Preliminary performance evaluation of a multilingual acoustic model for Ewondo, which allows us to outline paths for the construction of a robust model.

The rest of the paper is structured as follows. In Sect. 2 we briefly introduce and discuss the background related to this work. Section 3 presents our approach. In Sect. 4 we describe the experiments and discuss the results. Finally, we conclude in Sect. 4.

2   Background

2.1   Ewondo Language

Ewondo is a bantue language of central Cameroon, it is spoken by the Ewondo people in Cameroon, predominantly in the central and southern regions. It derived from the Fang-Beti language, which belong to the extensive Bantu language family, known for its diversity and widespread presence across sub-Saharan Africa.

The linguistic and cultural landscape of Ewondo is deeply rooted in the traditions and heritage of the Ewondo people. This language serves as a vital means of communication within the community, reflecting the rich history and social intricacies of its speakers. With its prevalence in urban areas, particularly in the capital city, Yaoundé, Ewondo plays a crucial role in daily interactions, commerce, and cultural expression.

The phonetics of Ewondo involve a set of distinctive consonants and vowels, contributing to its unique sound system. Pronunciation nuances, intonation patterns, and rhythmic elements are integral to conveying meaning accurately in spoken Ewondo. The language also incorporates a range of tones, a common feature in many Bantu languages, which further adds depth and complexity to its oral expression. In fact Ewondo is a tonal language, meaning that word meanings differ according to pitch, even if the consonants and vowels are the same [13] (Table 2 shows pairs of words of this type). The Ewondo language has 8 tones (Table 1), divided into punctual tones, which are tones for which the pitch remains invariable from the beginning to the end of the pronunciation, and modular tones, which vary in pitch.

Efforts to document and preserve Ewondo, both in written and oral forms, contribute to safeguarding the linguistic diversity of Cameroon. Like all Cameroonian languages, Ewondo uses the GACL[1] [14] alphabet (general alphabet of Cameroonian languages) based on the Latin alphabet. As with many endangered languages, Ewondo faces challenges such as globalization, urbanization, and the dominance of major languages. However, initiatives to promote language education, cultural exchange, and community engagement are crucial for

---

[1] https://www.silcam.org/fr/resources/archives/32295.

Table 1. Tones in Ewondo language

Pontuel tone		Modular tone	
Denomination	Notation	Denomination	Notation
Low[Tb]	[51v, v̌]		
High[HT]	[44v, v́]	High-Low[HLT]	[51v, v̂]
Medium[MT]	[33v, v̄]		
M-Low [MLT]	[ǀv]		
Supra-High[SHT]	[55v]	Low-High[LHT]	[15v, v̌]
Infra-LOW[SIL]	[12v]		

Table 2. Words that differ only in tone

Words	Translation	Words	Translation
minkud	*bag*	*minkúd*	*cloud*
zám	*raffia*	*zàm*	*good taste*
bám	*to scold*	*bam*	*to worry*
bóg	*to pil up*	*bog*	*to extract*
tag	*to rejoice*	*tág*	*to classify*

the continued vitality of Ewondo and its significance in the mosaic of Cameroon's linguistic heritage; this work is also in line with this aim. Despite of efforts and like all the languages of Cameroon, Ewondo remains a low resource language, i.e. numerical resources are almost non-existent. This constitutes a major difficulty for deep learning approaches to solving tasks such as speech recognition. However, recent approaches based on self-supervised models make it possible to tackle this type of language.

2.2   ASR with Self-supervised Models

Self-supervised learning is a machine learning paradigm where a model learns to make predictions about certain aspects of the input data without explicit supervision from labeled examples. First the NLP (Natural language processing) plume this approach has gained popularity for its ability to use large amounts of unlabeled data, often abundant in real-world scenarios. In fact, in self-supervised

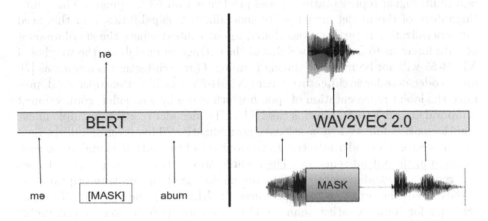

Fig. 1. Left. Example of a BERT pre training taok with the sentence "m n abum" (I'm pregnant), the word "n" is hidden and the model must predict it. **Rigth.** The same concept is applied to the audio signal, where certain portions are masked and wav2vec 2.0 must predict them.

learning, the learning algorithm creates its own supervision signal through a carefully designed pretext task. The pretext task is a task that is generated from the input data itself and doesn't require external annotations. The model is trained to solve this pretext task, and the acquired knowledge can then be transferred to downstream tasks where labeled data might be scarce.

The literature is replete with a number of self-supervised acoustic models (for the review of these model the reader can refers to [1]), but we have chosen to exploit the XLSR-53 a crosslingual version of wav2vec 2.0 [2] for its promising results on languages with small amounts of data. This model uses a pre-training task similar to BERT [10], illustrated in Fig. 1. This pre-training task consists of randomly masking words in sentences and asking the model to find the correct words. In the case of speech, parts of the signal are masked.

3   Wav2vec2.0 for Ewondo

We can divide our model in two parts; the cross-lingual speech representations (XLSR-53) [4] as a feature extractor and connectionist temporal classifier(CTC) [3] as a classifier. This section present the overall design of the model and different configurations used during experiments.

3.1   The Model

Our work is based on the Wav2Vec 2.0 [2] model. Overall, the Wav2vec 2.0 uses an auxiliary task similar to BERT [13], where certain parts of the signal are masked in order to be reconstructed by the system, it is trained by predicting speech units for masked parts of the audio. As shown in Fig. 3 we use as feature extractor the cross-lingual speech representations (XLSR-53) [4] version which is a multilingual representation model pre-trained on 53 languages. The multi-lingualism of the model increases its generalization capabilities, and this need for generalization is further exacerbated in our context where the small amount of data forces us to freeze the weights of the extraction model, i.e. the weights of XLSR-53 will not be modified during training. Our architecture is a same as [7], an encoder-decoder architecture where XLSR-53 acts like a encoder so it produce the latent representation of speech which is use by a decoder, connectionist temporal classifier (CTC) in this case. The CTC decoder model is a simple linear transformation followed by a softmax normalization. This layer should project output vector of encoder into the dimensionality of the output alphabet for each position in the output sequence. The main feature of this decoder is that it does not require strict alignment between the audio signal and its transcription, i.e. it only needs the input vectors (produced by XLSR-53) and the overall output sentence for training rather than a strict correspondence between input vector segments and output sentence segments. Let's take a closer look at the formal description of the fonctionnement of each part of our model.

Encoder. This XLSR-53 is a multilingual version of wav2vec 2.0 that consists of three parts: firstly, the feature encoder, which contains a multilayer convolutional neural network to process the raw waveform of audio speech. Secondly, the transformers, which are fed by the encoded feature and learn a contextualized representation from it, and thirdly the quantization module for selecting the speech unit to be learned from the latent representation space produced by the feature encoder. The purpose of this third part is to reduce the cardinality of the representation space and can be thought of as a function q that maps any vector x in the latent space to a vector $q(x)$ in a small group C of vectors called centroids. In a wav2vec, these quantized vectors are considered as the target of a transformer. As mentioned earlier, wav2vec uses a self-supervised strategy similar to BERT [10] for learning. This strategy involves randomly masking part of the feature encoder's output before sending it to the transformer, but the learning objective is formulated in a contrastive way and requires the identification of the correct representation, not of the encoded representation, but of the quantized latent audio representation q_t in a set of $K + 1$ quantized candidate representations $\tilde{q} \in Q_t$ which include q_t and K distractors for each masked time step. The lost contrastive function can be expressed as follows: $-log \frac{exp(sim(c_t, q_t))}{\sum_{\tilde{q} \sim Q_t} exp(sim(c_t, \tilde{q}))}$ where c_t is the transformer output, and $sim(a, b)$ represents the cosine similarity. This loss is augmented by a codebook diversity penalty to encourage the model to use all codebook entries.

To build a multilingual version of wav2vec 2.0, XLSR uses a shared quantization module on feature encoder representations, which means that feature encoder representations from different languages can be associated with the same quantized speech units. The multilingual quantized speech units produced by the quantization module are then used as targets for a transformer. This process forces the model to learn how to share discrete tokens between languages, creating a link between them that leads to a universalization of the acoustic representations obtained by the model.

Decoder. The CTC algorithm was developed by Grave and al. [3] for labeling sequence data task. As we previously said, it is alignment-free i.e in our case it doesn't require an alignment between the input vector segments produce by XLSR-53 and the output sentence segments. However, to get the probability of an output given an input, CTC works by summing over the probability of all possible alignments between the two. To define these possibles alignments, Grave et *al.* [3] introduce the ϵ symbol as a blank character in the output alphabet. This introduction solves two problems: (1) it is not logical to force each input step to align with an output in a speech recognition task; this symbol therefore marks a silence and (2) it marks the presence of several characters in a row, as it is difficult during recognition to know whether multiple identical letters in a row are a transcription of the same fragments or represent separate fragments, as is shown in Fig. 2 (a), putting an ϵ between them allows this difference to be made. As shown in Fig. 2 (a), a CTC alignment has the same length as the input, and we get the final output after merging the repeating characters and deleting the

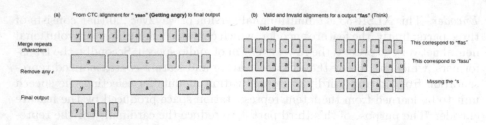

Fig. 2. (a) Steps taken by CTC to obtain the final transcription of the word "yaan" from one of its valid alignments. Firstly, we merge the repeating characters that are not interspersed with ϵ and secondly, we delete ϵ. (b) Examples of valid and invalid raw output for the word "fas". An alignment is valid when we can obtain a correct final transcription after the operation described in (a).

ϵ symbol. A CTC alignment is considered valid (Fig. 2 shows examples of valid and invalid CTC alignments for the "fas" output) for a given output if we can obtain the output from this alignment after the above-mentioned operations. CTC merges repeats characters between ϵ, so if an output has two of the same character in a row, then a valid alignment must have an ϵ between them. Based on previous description of alignment in CTC, during the training phase the objective is to maximize $P(Y|X) = \sum_{a \in A} \prod_{t=1}^{T} p(a_t|X)$ where a is a possible alignment and $p(a_t|X)$ is probability to have symbol a_t in time t in Y knowing X. $p(a_t|X)$ is given by the softmax at each time step. During inference phase CTC pick up $\hat{a} = argmax_Y(P(Y|a))$ as a final alignment and give an output after merging and remove operation.

As mentioned earlier in our model, XLSR-53 is frozen during the train process i.e. only the weights of decoder are modified during the process. Once the model is trained, if we would like to use it to find a likely transcription for a given new raw speech data (waveform), we proceed as follow: encoded it by XLSR-53 in a vector X, then CTC decoder tent to provide $\hat{Y} = argmax_Y(P(Y|X))$ where $P(Y|X)$ is the probability to have a sentence Y with X as input. Then greedy search is used as an inference process to pick up $\hat{Y}$, meaning we take the letter with the highest probability at each time step, until you receive the special token symbolizing the end.

3.2   Experiment Setup

We have chosen three main axes experiments, corresponding to different configurations of the features extractor model and data pre-processing.

Tokenization. If a token for speech recognition is the character, Rolando Coto-Solono's work [6] on Bribri (a Latin American language), has shown that it could be beneficial in a tonale low resources language context to make *tones*

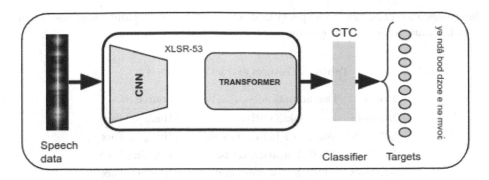

Fig. 3. ASR with self-supervised XLSR-53 Model. Speech data is passed in wavform to XLSR-53, which provides a vector representation of it. This representation is used by the CTC to predict a transcription.

explicit in the transcriptions of texts to be recognized. In fact, he proposes to introduce tones as explicit characters to be recognized. To verify this aspect, in ours experiments we introduced two tokenization principles presented in Table 3: TonSep where tones are explicit symbols to be recognized by the model, and ALL+tones where a tone was associated to a character and represented as one symbol to recognized.

Table 3. Differents types of tokenization

Type of tokenization	Example	Tokenization
ToneSep	ma wóg miǹtàg	m|a| w|´|o|g |m|i| |n|t|ˋ |a|g|
All+tones	ma wóg miǹtàg	m|a|w|ó|g|m|i|'|n|t|à|g|

Features Extractor. We have very little labelized data, so the XLSR-53 multilingual features extraction model is frozen, which means that it provides vectors from the weights derived from its pre-training. We propose to experiment with various XLRS-53 pre-trained models. These models are presented in the Table 4. The model named *XLSR-fb*[2] is the standard model [4], LeBench[3] is the Wav2vec 2.0 LeBenchmark [5] trained on data from the French language exclusively; the remaining models (XLSR-kw and XLSR-sw) being produced from XLSR-fb by fine turning on a specified language, in fact these models was built using standard model weights as initial weights, then pre-training was continued using unlabeled data of a specific language (kinyarwanda, swahili). Following this method, *XLSR-kw*[4] is a specialized XLSR-53 model for the Kinyarwanda

---

[2] https://huggingface.co/facebook/wav2vec2-large-xlsr-53.

[3] https://huggingface.co/LeBenchmark/wav2vec2-FR-7K-large.

[4] https://huggingface.co/lucio/wav2vec2-large-xlsr-kinyarwanda.

language and *XLSR-sw*[5] is a specialized XLSR-53 for the Swahili language, both of which are African bantue languages.

Table 4. Features extractions models

Model	Denomination	Source
XLSR-fb	Facebook XLSR-53	Hugging Face
LeBench	Wav2vec2 LeBenchmark	Hugging Face
XLSR-kw	XLSR kinyarwanda language	Hugging Face
XLSR-sw	XLSR for swahili language	Hugging Face

Language Model. Previous ASR models required both a language model and a pronunciation dictionary to transform classified fragment sequences of audio recordings into a coherent transcript. Recent end-to-end models have made this possible, but [2] has shown that the use of a language model in conjunction with wav2vec 2.0 significantly improves ASR performance, especially in low resources contexts. As part of our experiments, we tested the ASR model with the contribution of a bigram language model constructed from the transcriptions of the recordings in our dataset.

4   Experiments

The main objective of this work is to evaluate the performance of XLSR-53 for speech recognition of the Ewondo language. To achieve this goal, we collected and pre-processed speech data, then implemented the architecture described in Sect. 3. The literature has helped us choose the right tools to carry out these tasks. This section presents the details of these activities as well as the evaluation results.

4.1   Implementation Details

Dataset and preprocessing The Ewondo language has no public dataset for the ASR task, so we built a corpus from 103 sentences read by 5 speakers, including 4 men and one woman. We randomly selected 11 sentences for testing (2 min 30 s) and the remaining 92 sentences for training (21 min 51 s). The data was recorded at the computer science laboratory of Yaounde I, with a magnetophone, we use audacity[6] software for speech enhancement and artificially augmented these data with speechbrain [11] toolkit which using the method described in [8] whose idea is to introduce noise into data using simple transformation on a signal like *speed perturbation, frequency dropout, time dropout* to obtain new data.

---

[5] https://huggingface.co/Akashpb13/Swahili_xlsr.
[6] https://www.audacityteam.org.

Architecture. We used the extraction models from the hugging face repository[7] [12] as well as the recipes proposed on the same platform for the development of the ASR model[8]. The model hyperparameters are the same as [7]. We have used the KenLM [9] framework to build the bigram language model using transcript texts only; this model simple store the probabilities of word pairs appearing in the transcripts.

4.2   Results and Discussions

Tables 5 and 6 show performances of the ASR model according to the different extraction models, but also according to the use of the language model (LM/no.LM) during decoding, and the use of artificial data (sA/no.SA). In these two tables, we can see that the performance associated with *Lebench* is by far the worst of all configurations. This discrepancy can be explained by two facts: Lebench is a monolingual extractor trained only on French, a language linguistically distant from Ewondo. We can also see from these tables that the use of the language model systematically increases the performance of the ASR model, which is consistant with the results presented in [2]. We also note the counterintuitive results of artificial data augmentation, which degrades performance. One explanation for this contraction is to be found in the quantity of data from which the increase is made. Indeed, being of very small quantity, the artificial data acts as noise for the model.

Table 5. ASR model performance (%) with different feature extractors. Type of Tokenization = ALLfeat

	WER				CER			
	LM		no.LM		LM		no.LM	
	sA	no.sA	sA	no.sA	sA	no.sA	sA	no.sA
XLSR-fb	74.6	75.7	77.3	80.5	32.2	27.9	34.8	31.1
XLSR-rw	79.5	70.8	77.8	74.6	35.3	**28.6**	36.1	31
XLSR-sw	80	77.8	83.8	77.8	35.4	34.8	35.2	36.8
Lebench	97.3	97.3	100	100	93.9	97.3	100	100

Table 7 shows the average performance of the various ASR models in relation to the type of tokenization chosen. We can see that ToneSep is on average higher than ALL+tones, which means that it's better to recognize tones separately from characters in the low ressources case, a result in line with the recommendations of [6]. On average, the XLSR-FB standards perform best (70.8% on WER and 28% on CER), outperforming the specialized models, which can be explained by the richness of their representation, However, overall performance remains low

---

[7] https://huggingface.co.
[8] https://huggingface.co/blog/wav2vec2-with-ngram.

Table 6. ASR model performance (%) with different feature extractors. Type of Tokenization = ToneSep

	WER				CER			
	LM		no.LM		LM		no.LM	
	sA	no.sA	sA	no.sA	sA	no.sA	sA	no.sA
xlsr-fb	77.3	73	89.4	82	32.7	27.6	36.2	29.8
xlsr-rw	78.4	77.3	88.8	87.1	33.6	29.8	36.5	31.2
xlsr-sw	79.5	75.1	87.1	75.1	33.2	30.2	36.2	32.3
Lebench	97.3	97.3	100	100	93.9	97.3	100	100

Table 7. Average results for each tokenization methods

	WER				CER			
	LM		no.LM		LM		no.LM	
	sA	no.sA	sA	no.sA	sA	no.sA	sA	no.sA
ToneSep	82.8	80.4	84.7	83.2	49.2	47.1	51.5	49.7
ALLfeat	83.1	80.6	91.3	86	48.3	46.2	52.2	48.3

compared with the literature, which can be attributed to the extremely small amount of data available for training but also the distance existing between the target language and the languages underlying the pre-training of the acoustic model.

5   Conclusion

The aim of this paper was to apply the multilingual acoustic model wav2vec XLSR-53 to the Ewondo language for the transcription task. Preliminary results show overall poor performance compared to the literature in other languages (the best score being 70.8% on the WER and 28% on the CER). These results can be explain by the distance existing between the target language and the languages underlying the pre-training of the acoustic model. Although some similar work has already been carried out on African languages, our work reveals some singularities: firstly, the language of application, which to our knowledge is the first to be the subject of such a study; and secondly, the extremely small size of the dataset, which calls for greater finesse in pre-processing. In fact, in the literature working on low-resource data, datasets extend over at least several hours. This extremely low resource has enabled us to see the generalization limits of XLSR-53. Despite of the low performance, these experiments have enabled us to sketch out, apart from the need for additional data collection, some paths to follow in order to improve the transcription model. The first is to pre-train a multilingual wav2vec XLSR-53 model on Ewondo recordings, in order to familiarize the model with the language; the second is to pay particular attention to the explicitness of tones in transcription, which has proved beneficial to the model; the third is to build a more robust language model from a richer corpus of text. To further evaluate wav2vec in the Ewondo transcription task, a comparison with others features extractors models is a particularly interesting prospect.

Acknowledgments. This work has been funded by the European Union's Horizon 2020 research and innovation program under the Marie Skłodowska-Curie grant agreement No. 101007666, the Agency is not responsible for these results or use that may be made of the information.

Disclosure of Interests. The authors have no competing interests to declare that are relevant to the content of this article.

References

1. Mohamed, A., et al.: Self-supervised speech representation learning: a review. IEEE J. Sel. Topics Signal Process. **16**, 1179–1210 (2022)
2. Baevski, A., Zhou, H., Mohamed, A., Auli, M.: wav2vec 2.0: a framework for self-supervised learning of speech representations. CoRR, abs/2006.11477 (2020)
3. Graves, A., Fernández, S., Gomez, F., Schmidhuber, J.: Connectionist temporal classification: labelling unsegmented sequence data with recurrent neural networks. In: ICML 2006 - Proceedings of the 23rd International Conference on Machine Learning, pp. 369–376 (2006)
4. Alexis, C., Alexei, B., Ronan, C., Abdelrahman, M., Michael, A.: Unsupervised cross-lingual representation learning for speech recognition. arXiv (2020)
5. Evain, S., et al.: LeBenchmark: a reproducible framework for assessing self-supervised representation learning from speech. In: Interspeech 2021 (2021)
6. Coto-Solano, R.: Explicit tone transcription improves ASR performance in extremely low-resource languages: a case study in Bribri, pp. 173–184. Association for Computational Linguistics (2021)
7. Macaire, C., Didier, S., Benjamin, L., Emmanuel, S.: Automatic speech recognition and query by example for creole languages documentation. Assoc. Comput. Linguist. **2**(5), 2512–2520 (2022)
8. Heafield, K.: SpecAugment: a simple data augmentation method for automatic speech recognition. In: WMT@EMNLP (2011)
9. Park, D.S., et al.: KenLM: faster and smaller language model queries. In: Interspeech 2019 (2019)
10. Devlin, J., Chang, M.-W., Lee, K., Toutanova, K.: BERT: pre-training of deep bidirectional transformers for language understanding. arXiv (2019)
11. Ravanelli, M., et al.: SpeechBrain: a general-purpose speech toolkit (2021)
12. Thomas, W., et al.: Transformers: state-of-the-art natural language processing. In: Proceedings of the 2020 Conference on Empirical Methods in Natural Language Processing: System Demonstrations, pp. 38–45 (2019)
13. Bao, Z., Yip, M.: Tone. (Cambridge Textbooks in Linguistics), pp. xxxiv 341. Cambridge University Press, Cambridge (2002). Phonology, Cambridge University Presser (2023)
14. Essono, J.-M.: Langue et culture ewondo: par la grammaire, les textes et l'exercice: (suivi d'un lexique français-ewondo) (2012)

Self-supervised and Multilingual Learning Applied to the Wolof, Swahili and Fongbe

Prestilien Djionang Pindoh[1]([✉]) and Paulin Melatagia Yonta[1,2]

[1] Department of Computer Sciences, University of Yaoundé I, Yaoundé, Cameroon
prestilienpindoh@gmail.com
[2] IRD, UMMISCO, 93143 Bondy, France

Abstract. Under-resourced languages face significant challenges in speech recognition due to limited resources and data availability, hampering their development and usage. In this paper, we present a speech recognition model built upon existing frameworks based on self-supervised learning (Contrastive Predictive Coding (CPC), wav2vec and bidirectional version of CPC) by combining them with multilingual learning. This model is experimented on Wolof, Swahili, and Fongbe which are African languages. The results of our evaluation of representations on the automatic speech recognition task, using a similar architecture to DeepSpeech, highlight the model's capability to discriminate language-specific linguistic features, achieving a Word Error Rate (WER) of 61% for Fongbe, 72% for Wolof and 88% for Swahili.

Keywords: Self-supervised learning · Multilingual representation learning · Automatic speech recognition · Low endowed languages

1 Introduction

The technology of voice recognition has experienced significant advancements in recent years, offering numerous applications such as virtual assistants, transcription services, and voice command systems. However, these advancements have mainly benefited languages with abundant resources and extensive datasets, relegating under-resourced languages to the sidelines. These low-resourced languages, particularly African languages, face major challenges in developing accurate and efficient voice recognition systems due to limited data availability and the absence of dedicated linguistic models.

In recent years, self-supervised learning, in particular the contrastive learning approach [4] has emerged as a promising approach for learning representations from unlabeled data. Contrastive learning is a general framework that attempts to learn a feature space in order to bring together points that are related and discard points that are unrelated. Several methods exist today, such as Wav2Vec [2], CPC [1] and BCPC [3]. CPC represents an unsupervised machine learning approach that aims to derive meaningful, higher-level semantic representations from unprocessed data like text and audio. However, these methods also remain data-intensive for learning high-quality representations, which is a challenge for languages with limited resources. Kawakami et al. [3] use multilingual learning, which is an approach to machine learning that aims to learn a shared representation of speech from data coming from different languages. They learn representations with a large amount of multilingual data (With a high quantity of English) and then evaluate the transferability of these representations to other (sparsely endowed) languages, including Wolof, Swahili, and Fongbe, which yielded interesting results.

In this work, we aim to build a representation model specific to African languages capable of capturing the underlying features of each language. To this end, we use CPC, wav2vec, and bidirectional CPC in the context of multilingual learning. We use three low-resource languages: Wolof, Fongbe and Swahili. This allows us to have more data for training and to see how these methods manage to capture the unique characteristics of each language by evaluating them in the task. The underlying problem is to assess whether combining low-resource languages that share similar linguistic and phonetic characteristics improves the quality of features extracted for each language individually. This aims to enhance speech processing tasks.

The rest of this paper is structured as follows: we will first present the self-supervised learning approaches in our framework in Sect. 2, then we present the contrastive stacking model for multilingual learning with three African languages in Sect. 3. Section 4 is devoted to the experiments and the presentation of the results. We conclude this document in Sect. 5.

2 Related Work

Self-supervised learning is a machine learning method that allows an algorithm to learn from data without labels [9]. Self-supervised learning uses the structure of the data to generate its labels. This is different from unsupervised learning, which seeks to discover hidden patterns or structures in the data itself. This approach has seen growing interest in recent years thanks to the advent of new techniques, such as contrastive and generative learning.

Contrastive learning (Fig. 1) involves learning a metric space between two samples in which the distance between two positive samples is reduced while the distance between two negative samples is increased [4]. This technique can be used to learn representations useful for tasks such as image recognition or speech recognition. On the other hand, generative learning is a machine learning approach in which the model is trained to generate new data that resembles the training dataset. The generative model attempts to model the distribution of the training data so that new data can be sampled from this distribution. Generative models are often used for the generation of text, images, sounds, etc. In short, Generative Learning aims to generate new data that resembles the training dataset, while Contrastive Learning aims to learn discriminative representations by differentiating between similar and different pairs of data.

The InfoNCE metric is a commonly used measure in contrastive learning to assess the quality of learned representations. It is based on the principle

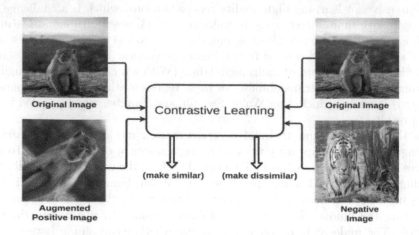

Fig. 1. Illustration of contrastive learning [9]

of maximizing the normalized mutual information between positive pairs and negative pairs.

The mathematical formula for the InfoNCE metric is as follows:

$$\text{InfoNCE} = \frac{1}{N} \sum_{i=1}^{N} \log \frac{\exp(\text{sim}(x_i, x_i^+))}{\exp(\text{sim}(x_i, x_i^+)) + \sum_{j=1}^{K} \exp(\text{sim}(x_i, x_i^{-j}))} \quad (1)$$

This cost function measures the probability that the positive example (x^+) is closer to the original example (x) than any negative example (x^-) (Fig. 1). The rationale behind this formulation is that the model should maximize the probability of selecting the positive among the negatives.

Aaron et al. [1] proposed contrastive predictive coding (CPC) (Fig. 2), a self-supervised machine learning method for extracting high-level semantic representations from raw data, such as text or sound. In the case of sound, this refers to the audio waveform of the signal, meaning the numerical features of the audio. It comprises two networks: an encoder that generates latent representations and an autoregressive network that generates contextual representations. The cost function used here is a mutual information (MI) lower bound called InfoNCE. This approach has produced impressive results for speaker identification and phoneme classification tasks. However, it was not used for speech recognition.

In general, CPC work as follows: given x a signal sliced into frames, the encoder network g_{enc} encodes the signal x at each time step t, yielding the latent representations z_t. Then the second autoregressive network g_{ar} produces the contextual representations c_t, taking into account the previous representations up to time step t, and predicts the future z_{t+k} from c_t, while maximizing the MI (Mutual Information) between c_t and the predicted z_{t+k}. Both representations,

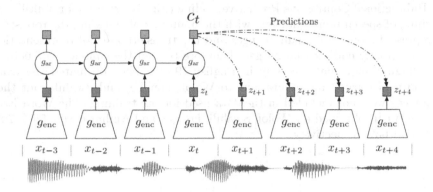

Fig. 2. Overview of CPC, the proposed representation learning approach. [1]

z_t and c_t, can be used as input for the automatic speech recognition model. In our case, we specifically use the contextual representations, c_t.

Schneider et al. [2] proposed Wav2Vec (Fig. 3), which is almost identical to CPC, but they use NCE (Noice Contrastive Estimation) as loss function instead of InfoNCE (a lower bound estimation of MI), and the encoder and context networks are made up of layers of causal convolutions, unlike CPC which uses convolutional networks on the encoder and a layer of GRU (Gated Recurrent Units) in the autoregressive network. Apart from this, they took the experiments further in the automatic speech recognition task. However, these experiments were conducted only in English and were not used in the context of low-resource languages.

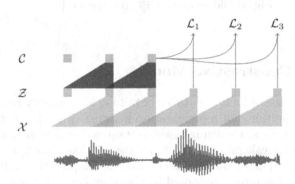

Fig. 3. Illustration of pre-training from audio data X which is encoded with two convolutional neural networks that are stacked on top of each other. The model is optimized to solve a next-time step prediction task [2]

Bidirectional Contrastive Predictive Coding (Fig. 4) was used for multilingual learning of speech representation, with the main aim of assessing the robustness of representations to domain changes and the transferability of representations to other poorly endowed languages. They pre-trained their model on 8,000 h of multilingual audio data, mainly in English (95%). They evaluated the transferability of learned representations to Wolof, Fongbe, and Swahili, but these languages were not included in the data used for pre-training. The Word Error Rates (WER) obtained for Wolof, Swahili, Fongbe and Amharic were 55%, 70%, 57%, and 65% respectively.

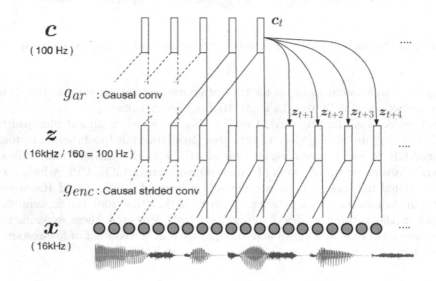

Fig. 4. Bidirectional CPC Illustration [3]

3   Stacking Constrastive Models

To overcome the issue of limited data for obtaining high-quality representations, we implemented multilingual representation learning (Fig. 4). For this purpose, we used the self-supervised learning methods proposed by Aaron et al. [1], Schneider et al. [2], and Kawakami et al. [3]. Since the CPC and wav2vec models were not originally trained in a multilingual context, we adopted the multilingual representation learning approach proposed by Kawakami et al. [3], which we applied to all three methods before using them for subsequent tasks.

We used speech recognition datasets in three African languages, collected as part of the ALFFA project[1]: Fongbe (A. A Laleye et al. [8]), Swahili (Gelas et al.

---
[1] http://alffa.imag.fr.

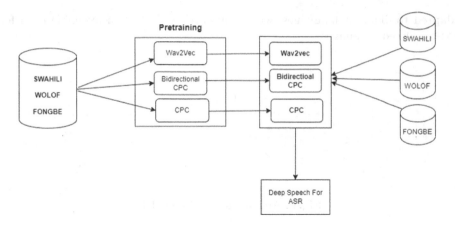

Fig. 5. Architecture of the model

[6]), and Wolof (Gauthier et al. [5]). These languages are characterized by unique phonological properties such as pitch harmony and distinct phonetic inventories. It is worth noting that these African languages have limited resources available, with less than 20 h of transcribed speech data for each of them.

Once we obtained these datasets, we followed the following steps for representation learning as shown in Fig. 4:

- The data was preprocessed and normalized to reduce noise and mitigate volume differences between the data. Preprocessing involves applying filters to reduce background noise and normalize volume across recordings. This step ensures consistency in input data for the learning model.
- We used the data set to train the representation learning algorithms, namely CPC, wav2vec, and bidirectional CPC. We thus obtained three pre-trained models in all languages.
- To predict representations for a single language, we fed the data from that language into one of the pre-trained models. The predicted representation vectors were then used to train the automatic speech recognition (ASR) model.

Contrastive learning, using a mixture of audio from several languages, presents an interesting opportunity for multilingual speech representation learning. This approach is based on the principle that, according to the contrastive learning methodology, positive sampling is independent of a specific language.

By combining data from different languages during contrastive learning (Fig. 5), the model is exposed to a variety of acoustic and linguistic structures, which promotes the emergence of general and shared representations. As a result, the model can learn to extract speech-relevant features from different languages, while capturing the similarities and differences between them.

In conclusion, the use of contrastive learning with a mixture of multilingual data offers a promising approach to develop speech representations that are adapted to different languages, while maximizing the use of available data for low-resourced languages.

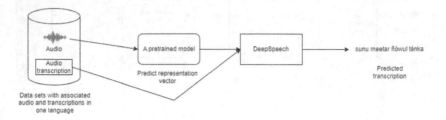

Fig. 6. Architecture of the model

Once the representations are learned, we use the obtained models to predict the representation vectors of the data (Fig. 6), which will be used as inputs to the speech recognition algorithm DeepSpeech2 Small (a reduced version of DeepSpeech2 [7]). DeepSpeech2 is an automatic speech recognition model based on deep learning. It uses a recurrent neural network (RNN) with long short-term memory (LSTM) to perform the task of transcribing speech into text. It uses the spectrograms (numerical representation vector) as model inputs but in our case we have replaced the spectrograms by the representation vectors predicted by the pre-trained models.

During the training phase of the ASR model, the parameters of the representation models were kept constant, while the parameters of the speech recognition models were supervised and trained using one dataset at a time [3]. The models were evaluated using the standard Word Error Rate (WER) metric on held-out test data. The WER is calculated as the number of words incorrectly predicted divided by the total number of words. It measures the accuracy of the model's transcription, where a lower value indicates better performance of the automatic speech recognition model. To evaluate the impact of multilingual learning, we also trained monolingual representation learning models to formulate hypotheses and compare the results.

4 Experiments and Results

ALFFA (African Languages in the Field: Speech Fundamentals and Automation) is a research project aimed at collecting high-quality linguistic data for understudied African languages using speech recognition to enable speakers of these languages to access information in their language. The ALFFA project has gathered data for several African languages, including Wolof, Swahili, and Fongbe, available on GitHub[2]. These languages are considered low-resource, hence the inherent interest in studying them. Moreover, from a more objective standpoint,

---

[2] ALFFA: https://github.com/besacier/ALFFA_PUBLIC.

these languages were chosen because there existed significant and usable public data sources for each. The Wolof language, primarily spoken in Senegal, is represented in a dataset featuring 21 h of recorded speech from 18 distinct speakers. Wolof distinguishes itself with an extensive vowel system and a unique possession indicator, "ñu." Noteworthy features include its nominal class system and the use of specific prefixes based on these classes, contributing to Wolof's linguistic distinctiveness. This dataset is meticulously curated, drawing from diverse sources such as proverbs, narratives by Kesteloot and Dieng (1989), transcriptions of healer debates, a song titled "Baay de Ouza," and two dictionaries: "Dictionnary wolof-french" and "Dictionnary french-wolof." Additionally, data from the Bible, Wikipedia, and the Universal Declaration of Human Rights enrich the corpus. Fongbe, also known as Fon, is predominantly spoken in Benin, where it holds the status of a national language. It boasts a tonal system influencing word meaning through varying pitch. The Fongbe dataset consists of 9 h of recorded speech from 29 different speakers. The textual corpus content includes primarily biblical texts, a variety of texts related to daily life, the Universal Declaration of Human Rights, texts on education, songs, and folktales. Swahili, spoken primarily in East Africa, serves as the national and official language in both Kenya and Tanzania. It extends its influence to countries like Uganda, Rwanda, Burundi, and parts of the Democratic Republic of the Congo. As a Bantu language, Swahili is characterized by an agglutinative structure, incorporating prefixes and suffixes for grammatical nuances. The Swahili dataset consists of 12 h of recorded speech from various speakers, representing diverse socio-economic and educational backgrounds. Recordings originate from web-based news broadcasts and Swahili text extracted from informational websites. Manual transcriptions are available for each recording.

We used an unofficial implementation from GitHub[3] for the CPC model that we modified for our application case and the official implementation[4] of the Wav2Vec model for our experiments. Since no bidirectional CPC model implementation was available, we modified the code of the Wav2Vec model to make it bidirectional[5] (as mentioned in the article by Kawakami et al. [3]). Similarly for DeepSpeech, we used an implementation available on the Keras framework site[6] that we adapted to use the numerical vectors predicted by our pre-trained models rather than spectrograms.

4.1 Descriptions and Parameters of the Representation and ASR Learning Algorithms

The parameters of the algorithms we used to learn the representations are shown in Table 1.

---

[3] https://github.com/jefflai108/Contrastive-Predictive-Coding-PyTorch.

[4] https://github.com/facebookresearch/fairseq/tree/main/examples/wav2vec.

[5] https://drive.google.com/drive/folders/1IWhFaQzTPADUJo0j8tnSzorvcFaUloji? usp=sharing.

[6] https://keras.io/examples/audio/ctc_asr/.

Table 1. Parameters of the models

	CPC	BCPC	Wav2vec
Number of negative samplings	10	10	10
Timestep	12	12	12
Audio window length (frames)	20480	150000	150000
Optimizer	Adam	Adam	Adam
Initial learning rate	0.0004	0.0001 with gradient clipping, maximum norm of 5.0	0.0001
Batch size	8	128	8
Loss	InfoNCE	InfoNCE (sum of backward and forward)	InfoNCE
Encoder size	512	512	512
Encoder Layers	convolutional layers with kernel size (10, 8, 4, 4, 4) and strides (5, 4, 2, 2, 2).	5 causal convolutional layers with kernel size (10, 8, 4, 4, 4, 1, 1) and strides (5, 4, 2, 2, 2, 1, 1).	5 causal convolutional layers with kernel size (10, 8, 4, 4, 4) and strides (5, 4, 2, 2, 2).
Decoder size	256	(fwd: 256, bwd: 256): concatenation (c = [fwd, bwd] => 512)	512
Decoder Layer	An unidirectional GRU layer	13 causal convolutional layers with kernel size 1, 2, ..., 13 and stride 1	9 causal convolutional layers with kernel size 3 and stride 1

For speech recognition, DeepSpeech2 Small. The features of this model are as follows: the model features two 2d-convolutions with kernel sizes (11, 41) and (11, 21) and stride sizes (2, 2) and (1, 2), as well as a one-way recurrent neural network (GRU) above the output of the convolution layers. A linear transformation and a softmax function are applied to predict frame-level character probabilities. Training is performed using a batch size of 8 and a learning rate of 0.0001.

We use WER, which is a popular metric for evaluating the performance of automatic speech recognition algorithms because it takes into account errors such as word substitution, insertion, and deletion, which are common types of errors in speech recognition. A lower WER value indicates better performance, as it means fewer errors in the output generated by the speech recognition system, and its value is expressed as a percentage.

4.2 Monolingual Learning

The results of the experiments we conducted using a single language to learn the representations and then using them to train the automatic speech recognition model are presented in Table 2. The metric used is the Word Error Rate (WER) and is expressed as a percentage.

Table 2. Results on monolingual learning

	CPC	BCPC	Wav2vec	Kawakami et al
WOLOF	**80**	85.7	80.1	**55**
FONGBE	83	81	**68**	**57**
SWAHILI	96	95	**90**	**70**

4.3   Multilingual Learning

The results obtained using the learned representations on all of these datasets for the downstream task (trained with the previous parameters) are as the Table 3:

Table 3. Results on multilingual learning

	CPC	BCPC	Wav2vec	Kawakami et al.
WOLOF	78.9	75.2	**72.6**	**55**
FONGBE	82	77.5	**61**	**57**
SWAHILI	95	93	**88**	**70**

In the results obtained, we can see a clear improvement in WER in the case of multilingual learning compared to monolingual learning. This suggests that the number of datasets used to train the representation model had an impact on the quality of the extracted representations and that the inclusion of other languages improved the features extracted from the input audio signal, resulting in improved performance in the downstream task.

We observed that all these approaches yielded promising results, especially wav2vec, which enabled us to create high-quality speech representations. However, these results are still far from state-of-the-art. This can be explained by the fact that the models we used for pre-training require a large amount of data to learn good-quality representations and that they failed to correctly capture some language-specific speech features, despite having combined several data sources (42 h in total).

These results show that the pretraining of the small amounts of available data, pooled in multilingual models, can allow to have a robust model that captures the acoustic characteristics of each language.

5   Conclusion

This paper shows that pre-training on small amounts of available data, pooled in multilingual models, can allow having a robust model that captures the acoustic characteristics of each language. Thus, one could build a single model for several languages instead of building a different model for each language.

Although our work has made significant progress in self-supervised learning of speech representations for under-resourced African languages, there are still many perspectives to explore in order to further improve the results.

First, it would be interesting to expand the training data corpus for representation learning models by collecting more speech data for a broader set of African languages. This would make it possible to obtain even more diverse speech representations tailored to a wider range of under-resourced languages.

Next, the speech recognition models, especially DeepSpeech, could be enhanced by developing language models specific to each language. By incorporating language-specific knowledge, it would be possible to achieve more accurate and higher-quality transcriptions.

In addition, new multilingual self-supervised learning techniques are worth exploring, such as using generative methods based on variational autoencoders or adversarial generative networks.

By combining these different avenues for improvement, we hope within the next few years to have accurate speech recognition systems for a wide variety of under-resourced African languages. This will help strengthen the presence of these languages in speech technologies and enable their speakers to more easily access information in their own language.

References

1. van der oord, A., et al.: Representation Learning with Contrastive Predictive Coding. arXiv:1807.03748v2 (2019)
2. Schneider, S., Baevski, A., Collobert, R., Auli, M.: Wav2vec: unsupervised pre-training for speech recognition. arXiv:1904.05862v4 (2019)
3. Kawakami, K., Wang, L., Dyer, C., Blunsom, P., van den Oord, A.: Learning Robust and Multilingual Speech Representations. arXiv:1904.05862v4 (2020)
4. Arora, S., Khandeparkar, H., Khodak, M., Plevrakis, O., Saunshi, N.: A theoretical analysis of contrastive unsupervised representation learning. In: PMLR 1997 (2019)
5. Gauthier, E., Besacier, L., Auli, M., et al: Collecting resources in sub-saharan African languages for automatic speech recognition: a case study of Wolof. In: LREC (2016)
6. Gelas, H., Besacier, L., Pellegrino, F., Auli, M.: Developments of Swahili resources for an automatic speech recognition system SLTU . In: Workshop on Spoken Language Technologies for Under-Resourced Languages, Cape-Town, Afrique Du Sud (2012)
7. Amodei, D. et al.: Deep speech 2: end-to-end speech recognition in English and Mandarin In: International Conference on Machine Learning (2016)
8. Laleye, A.A., et al.: First automatic Fongbe continuous speech recognition system: development of acoustic models and language models. In: Federated Conference on Computer Science and Information Systems (2016)

9. Jaiswal, A., Ramesh Babu, A., Zaki Zadeh, M., Banerjee, D., Makedon, F.: Survey on contrastive self-supervised learning. Technologies **9**(1), 2 (2020)
10. Bengio, Y., Courville, A., Vincent, P.: Representation learning: a review and new perspectives. IEEE Trans. Pattern Anal. Mach. Intell. **35**(8), 1798–1828 (2013)

Explaining Meta-learner's Predictions: Case of Corporate CO2 Emissions

Ingrid Pamela Nguemkam Tebou[1,2]([✉]), Norbert Tsopze[1,2][iD],
and Dieudonné Tchuente[3][iD]

[1] University of Yaoundé 1, Yaoundé, Cameroon
[2] Sorbonne University, IRD, UMMISCO, 93143 Bondy, France
ingrid.nguemkam@facsciences-uy1.cm
[3] TBS Business School, Toulouse, France
d.tchuente@tbs-education.fr

Abstract. Production activities of companies very often leads to release of carbon dioxide (CO_2) into the atmosphere which pollutes the environment. It is therefore necessary for companies to control and reduce their pollution levels, and this requires knowing the amount of CO_2 emissions that can be produced and identifying the factors responsible for it. Several works have been carried out with the aim of predicting the quantity of CO_2 emitted at the company level. One of these works use a Stacked Generalization ensemble model for prediction, and get better performance. This aforementioned work proposes an analysis of the role of variables in the prediction, but these explanations are provided based on some weak learners while the best prediction performance is obtained by the whole ensemble model. In this work we propose to explain the prediction of CO_2 emissions from the whole Stacked Generalization ensemble model itself. For this we propose an explainability method that combines the strenghts of two state of the art explainability techniques (LRP and SHAP). Experiments show that the proposed method can be relevant both in term of stability and fidelity.

Keywords: CO_2 emissions · Explainable Artificial Intelligence · Stacked Generalization

1 Introduction

Global warming is currently one of the major societal challenges on a planetary scale, for which one of the major causes is environmental pollution by greenhouse gases. Since carbon dioxide (CO_2) is the most pollutant of all greenhouse gases, reducing carbon dioxide emissions is of vital importance (National Geography Society[1]). Global carbon dioxide (CO_2) emissions from energy combustion and industrial processes increased by 0.9%, or 321 Mt, in 2022, reaching a new all-time high of 36.8 Gt. (IEA CO_2 emissions report[2]). This new record reinforces

---

[1] https://education.nationalgeographic.org/resource/greenhouse-effect/, last accessed 2023/07/02.

[2] https://www.iea.org/reports/co2-emissions-in-2022, last accessed 2023/07/02.

the need for the industrial sector and therefore corporate to work harder in order to reduce CO_2 emissions, and for regulators in the field to increase the control of CO_2 emissions from corporates. Whether controlling or reducing CO_2 emissions, this requires knowing the amount of emissions that can be produced by companies and the factors responsible for the amount of CO_2 emitted.

Artificial Intelligence and more particularly the branch of Explainable Artificial Intelligence can be used to identify the factors responsible for the amount of CO_2 emitted. Several studies have focused on estimating CO_2 emissions generated by companies. Some work predict CO_2 emissions in a geographical area (city, country, etc.) or for specific industry sectors using a single machine learning model. The work of Nguyen et al. [8] uses an ensemble learning model consisting of a meta-learner and six weak learners, to predict the amount of CO_2 emissions at the level of companies and get better prediction performance than others. In that work, they also propose an analysis of the role of the variables in the prediction, but this analysis is based only on certain weak learners (of the linear regression type and decision trees) whereas the best prediction performances are obtained by the final prediction of the meta-learner. This implies that in case of the missed classification by the weak learner, the explanation is done on the false output. This poses the problem of the fidelity of the output explanations. Ensemble learning models can be explained using agnostic explainability methods such as SHAP [6] and LIME [9], except that these methods view the ensemble model as a black box and thus explanations are provided regardless of the proper functioning of ensemble learning model type. We propose in this work an explainability method of a Stacked Generalization model which takes into account its internal organization. The idea is to calculate the contribution of each weak learner on the meta-learner output, and from these contributions, deduct those of different input variables. Our contribution covers three aspects. Firstly, several works [4,5,10,11,14,15], focus their attention on predicting CO_2 emissions without explanations. For those which explain/analyze carbon emissions predictions, they rely on intrinsically interpretable models [8]. In this work we rather propose a more generic post hoc method. Secondly, the proposed method combines two explainability methods widely used in the literature (LRP and SHAP). SHAP method is model agnostic and widely used to explain all types of models, while LRP method is widely used to explain neural networks. We combine these two methods into a single one, taking advantage of the contribution conservation property of LRP, and the fair distribution property of SHAP. Lastly, the proposed method is adapted to explain predictions of Stacked Generalization Ensemble models.

The paper is organized as follows: Sect. 2 presents existing works on the prediction of CO_2 emissions and existing explainability methods, then Sect. 3 presents the method we propose, then Sect. 4 a summary of the experimentation carried out and the results obtained and, finally, Sect. 5 concludes our work.

2 Related Works

We will talk about existing work on CO_2 emissions prediction and explainability methods. Work on predicting corporate carbon emissions remains relatively recent [8]. Most of these works use a basic machine learning model among a wide range of algorithms such as linear regressions [4,5]; neural networks [10,14]; support vector machines [11,15]. All these works focus on carbon footprint prediction and the data manipulated is restricted to either a sector of activity, a city or a country, as well as the characteristics and the number of companies taken into account for the prediction are not numerous. In 2021, [8] propose to use a Stacked Generalization ensemble model to predict CO_2 emissions which have better performance and an explanation of predictions is also presented but only based on linear models and decision trees predictions, however, it is the meta-learner who produces the best prediction performance.

Machine learning algorithms produced impressive performance in the process of decision making but the logic followed to obtain these results is sometimes difficult to understand due to the complexity of these algorithms. Methods have been developed to explain the results provided by a machine learning model. Following the dependency of the method on the model, the literature distinguishes two types of explainability methods: agnostic explainability methods and model-specific explainability methods [1]. Agnostic methods are those that can be applied to all types of Machine Learning algorithms, while model-specific methods only apply to a single class of algorithms. Local Interpretable Model-agnostic Explanations [9] and SHapley Additive exPlanations [6] are agnostic methods.

LIME aims to explain the prediction of an instance by approximating it locally with a human-understandable model. The principle is to simulate the behavior of the complex model by using a surrogate model in the neighborhood of the instance to be explained. The surrogate model used should be simple and easily understandable such as linear regression and decision tree. One of the limits of the LIME method is that the quality of the explanation provided strongly depends on the choice of neighbors to take into account when fitting the surrogate model, if this choice is not good the explanation will be bad quality. Another limit is that the choice of neighbors being made randomly at each execution of the algorithm, this produces different explanations which makes this method unstable.

SHAP is a unification of several local explainability methods. It is based on the value of Shapley resulting from the theory of cooperative games. The SHAP method transposes game theory into machine learning where the game is an observation, the players are the characteristics, the total gain is the prediction; thus the Shapley value will correspond to the contribution of a feature for a given prediction. The main limitation of SHAP is the computation time needed to determine Shapley values, especially for a dataset with a large number of features. In practice, we try to calculate a value closer to the exact value by choosing certain combinations instead of all combinations.

Among the Model-specific methods, we have the Layer-wise Relevance Propagation method [2]. LRP is an explainability method designed for neural networks, which identifies input layer neurons that are decisive for prediction. The principle is to redistribute the prediction $f(x)$ backwards in the neural network, starting from the output layer of the network and back-propagating to the input layer. The propagation procedure implemented by LRP is subject to a conservation property, where what has been received by a neuron must be redistributed to the lower layer in equal quantity. This propagation is done by means of specially designed local propagation rules. LRP is subject to the conservation property during back-propagation which makes it a stable and robust method, in addition to being based on a mathematical theory (Deep Taylor Decomposition). But LRP only works on Neural Network models.

3 Explaining Meta-Learner's Predictions

The method we propose aims to explain the predictions of a meta-learner. The meta-learner is used to combine the predictions produced by different weak learners, by using a machine learning model which takes as input the predictions of the weak learners and makes a combination of these to produce the final prediction. Figure 1 illustrates the meta-learner architecture.

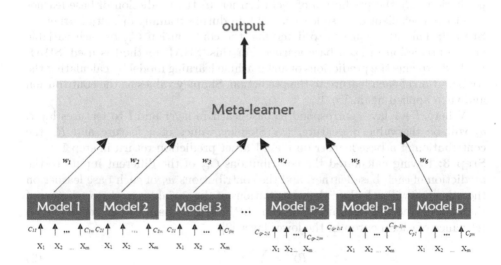

Fig. 1. Meta-learner Architecture

This architecture can be assimilated to the one of a neural network in that the features would be seen as input layer, and the set of weak learners would be seen as a hidden layer in which each weak learner could be considered as a neuron and the final prediction of the meta-learner would be seen as the output layer consisting of a single neuron producing the final result.

The idea is to exploit this similarity to put together two explainability methods widely used in the literature : LRP (Layer-wise Relevance Propagation) [2] and SHAP (SHapley Additive exPlanation) [6]. SHAP is used to determine the contributions of the variables and LRP to take into account the operation similar to the neural network of a meta-learner. For this we propose a three-step method :

1: Calculate the contributions of each weak learner on the prediction of the meta-learner using the LRP method,
2: Calculate the contributions of each input of each weak learner using the SHAP method,
3: Aggregate contributions using the LRP method.

Step 1: This step consists in determining the relevance w_i of each weak learner. These relevances will correspond to the contributions of each base learner to the meta-learner's prediction (final output). We use for this the LRP-0 propagation rule [2] which allows us to obtain using Eq. 1 the contributions of each neuron in layer l from the relevance of the neuron in the previous layer $l + 1$:

$$R_{i \leftarrow k}^{(l,l+1)} = R_k^{(l+1)} \cdot \frac{a_i w_{ik}}{\sum_h a_h w_{hk}} \tag{1}$$

where, $l + 1$ layer correspond to the meta-learner output layer and l to base learners layer. $R_{i \leftarrow k}$ will be the contribution of base learner i on meta-learner prediction , R_k the prediction of meta-learner, a_i the prediction of base learner i and w_{ik} coefficient of base learner i obtain during training of meta-learner.

Step 2: This step consists to determine the contribution C_{ij} of each variable on the prediction of each base learner. For this, SHAP method is used. SHAP method explains the predictions of any machine learning model by calculating the contribution of each feature to the prediction, Shapley values as the contribution and then applied again Eq. 1.

Where, $l + 1$ layer correspond to base learners layer and l to features layer. a_i will be the value of feature, w_{ik} Shapley values of a feature and R_k the contribution of a base learner on meta-learner prediction return in step 1.

Step 3: Having calculated the contributions C_{ij} of the different inputs to the prediction of each base learner, and the contributions w_i of each base learner on the other, we will calculate the contributions of the variables to the prediction of the meta-learner. To do this, we use the LRP formula used to determine overall relevance of each neuron in the lower layer [2]:

$$R_i^{(l)} = \sum_k R_{i \leftarrow k}^{(l,l+1)} \tag{2}$$

where, R_i is the contribution of a variable i on meta-learner prediction, $R_{i \leftarrow k}$ is the contribution of a variable i on a prediction of a base learner k.

The proposed method retains the same computational difficulties and the instability of the contributions of the Shap method. It also retains the specificity of only applying to models with an architecture similar to neural networks such as the LRP method. The method is positioned in the category of model-specific explainability methods, since it can be used on Stacked generalization ensemble learning models using a linear regression as meta-learner. The method can be used for local explainability but can be generalized by using the average of the obtained contributions to provide a global explanation of the model predictions.

4 Experiments and Results

4.1 Dataset

We used Thomson Reuters Full ESG dataset provided by Thompson Reuters Corporation which contains data of CO_2 emissions for several firms around the World. The dataset contains 14,531 observations and 113 numerical and categorical features with 614,100 missing values. As the dataset had a high rate of missing values, features with at most 5,267 missing values were retained, then categorical features were removed, and missing values replacement was done using Multiple Imputation by Chained Equations (MICE) algorithm [12]. We also normalized the data by applying standardization on all numeric features:

$$x = \frac{x - \overline{X}}{\sigma} \tag{3}$$

A dataset of 14,531 observations and 53 variables was obtained at the end.

4.2 Implementation

We used stacked generalization ensemble model for predicting the carbon footprint of companies following the methodology of Nguyen et al. [8]. The model is formed by six weak learners: Ordinary Least Squares, Elastic Net, Random Forest, and XGBoost, K-nearest Neighbors and Multi-Layers Perceptron. Weak learners predictions are combined using two strategies: average prediction and meta-learner (OLS, Elastic Net). Hyperparameter of weak learners was done using HyperOpt and GridSearch python libraries.

For explainability, we used the agnostic variant KernelShap of Shapley Additive exPlanations (SHAP) method [6] to explain predictions of all models. KernelShap is available with SHAP python library.

4.3 Evaluation

We used K-fold Cross-validation (with k=10) for hyperparameter tuning, only the training set was used. The evaluation metrics used are: Mean Absolute Error (MAE) and coefficient of determination (R^2). MAE is used to observe difference between real values and predictions and R^2 is used to evaluate the quality of regression.

$$MAE = \frac{1}{n} \sum_{i=1}^{n} |y_i - \tilde{y}_i| \tag{4}$$

$$R^2 = 1 - \frac{\sum(y_i - \tilde{y}_i)^2}{\sum(y_i - \overline{y})^2} \tag{5}$$

Explainability methods were evaluated using fidelity and stability metrics. Fidelity was evaluated following the approach proposed by [7]. This approach consists of identifying the most important variables by removing them, and observe the change on predictions. Stability was evaluated using the Variable Stability Index (VSI) [13]. Variable Stability Index measures whether the same explanatory variables recur across repeated executions.

4.4 Results

Prediction Performance. The prediction performances obtained by weak learners are summarized in Table 1.

Table 1. Performance of weak Learners

	OLS	Elastic Net	KNN	Random Forest	XGBoost	MLP
MAE	0.356	0.301	**0.099**	0.123	0.131	0.127
R^2	0.173	0.129	0.831	0.755	0.818	**0.835**

From Table 1, we can see that KNN model, outperforms other weak learners with the smallest MAE (0.099) which shows that the model predicts more better. The Multi-Layer Perceptron has the highest coefficient of determination, although it does not have the lowest error. Decision trees predict well. Linear regressions, on the other hand, performed poorly with the highest MAEs and the predictions mean that they did not follow the distribution of the data at all. We test three differents combinations strategies of weak learners for prediction, Table 2 summarized the results.

Table 2. Performance of combination strategies

	Mean-based	Meta-Elastic Net	Meta-OLS
MAE	0.161	0.158	**0.116**
R^2	0.758	0.739	**0.856**

This Table shows that the best strategy for combining the predictions of the base learners is the meta-learner Ordinary Least Squares linear regression with MAE of 0.116 and a coefficient of determination of 0.856. This errror value is close to the one of KNN base learner and this R^2 is the best value of all the base learners models. Nguyen et al. [8] obtained MAEs within the interval [0.694; 1.133], whether for weak learners or for meta-models. Although the dataset is not the same, we note an improvement in prediction performance in our case.

Explainability. SHAP method was tested on the meta-learner and on all the weak learners to determine the contributions of the variables for each of these models. We compared the top-5 of most important features obtained for each model Fig. 2 presents the contribution of features in each top-5 and the contributions of these features in the other models.

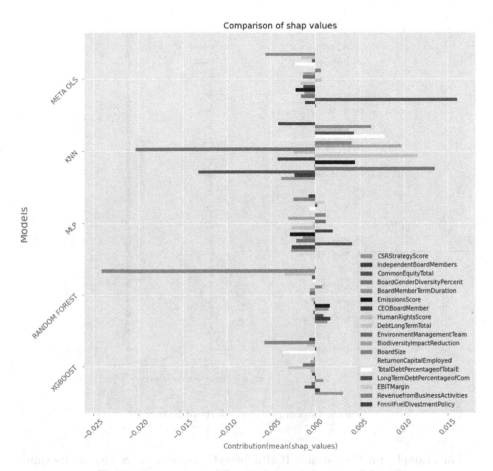

Fig. 2. Comparison of Top-5 of most important features

Each top-5 of most important features differ from one model to another but with some features such as RevenuesfromBusinessActivities, CommonEquityTotal, EBITMargin, which always end up in the top-5 regardless of the model. The others features have a positive contribution in a model but negative in another and vice-versa. From these observations, it would be incorrect to choose certain base learners to explain the predictions of the meta-Learner because these explanations could be biased. We opt to explain the ensemble learning model itself.

We will work with the contributions of the meta-learner OLS which has the best performance. Figure 3 present feature importance combined with feature effects on prediction. Each point on the summary plot is a Shapley value for a feature and an instance. The position on the y-axis is determined by the feature and on the x-axis by the Shapley value.

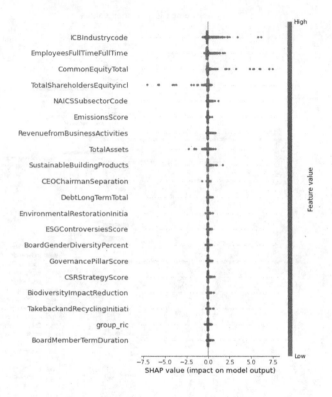

Fig. 3. Importance and effects of features

For example, for the variable ICBIndustryCode, we can see that as its value increases, its Shapley value (and therefore its impact on prediction) increases also. On the other hand, for the TotalShareholdersEquityIncl feature greater its value, more its Shapley value decreases.

SHAP method perceives the Stacked Generalization model as a black box model, and explains predictions without taking into account the internal structure of this model. We think that taking into account the internal workings of Stacked Generalization model could improve the explanation of predictions. We therefore calculate the contributions of each feature with the proposed method, and then evaluate the fidelity and stability of proposed method in comparison with SHAP. First, the average fidelity is calculated on test set. The average fidelity is obtained by averaging the individual fidelity scores of each observation in the test set. Table 3 shows the average of fidelity scores of each observation in test set, obtained from contributions of variables for SHAP and the proposed method.

Table 3. Average fidelity of SHAP and LSMeta

	SHAP	LSMeta
Average fidelity	0.216 (0.742)	0.096 (0.256)

We note that the average fidelity score obtained for the two methods are different and non-zero, which means that both methods produce faithful explanations of the model output, but each at a different level of fidelity. The average of the fidelity scores of the proposed method is lower than that of SHAP but with a lower standard deviation. The standard deviation values show that the mean of fidelity scores with LSMeta is closer to zero, indicating that the data points are more close to the mean, whereas with SHAP, the data points are further from the mean. It is therefore not suitable to compare directly on the basis of average fidelity scores, as the mean do not represents the two distributions in the same way.

Next, we look at the individual fidelity scores of each observation for the two methods. We compare the percentage of observations on which the LSMeta method has a higher fidelity than SHAP and vice-versa. Table 4 shows the proportions of the test set in which each method has a higher fidelity score than the other.

Table 4. Proportion of individual fidelity scores

	SHAP	LSMeta
% Higher Fidelity	59%	41%

We observe that on 59% of observations, SHAP provides explanation with higher fidelity score than LSMeta, while on 41% of observations its rather LSMeta which provides higher fidelity. We believe that these 41% are interesting results which already show that the proposed method can provide more faithful explanations than the SHAP method in some cases. The results obtained so far only take into account five important variables to be returned by the explainability methods. To go further in our experiments, we varied the size of the explanation (the number of important variables) to be returned by SHAP and LSMeta methods and then observed the evolution of the fidelity and stability on these different explanation sizes. To do this, we chose to return as important variables, one variable, then $\frac{1}{10}$th of the total number of variables, then $\frac{1}{4}$, $\frac{1}{3}$, and $\frac{1}{2}$. Table 5[3] represents the results obtained for the different sizes of explanations. Figures 4 and 5 illustrate the results of Table 5.

Table 5. Comparison of fidelity and stability scores

	1	1/10	1/4	1/3	1/2
Fidelity LSMeta >=SHAP (%)	34.25	41.07	55.62	58.17	65.67
Stability LSMeta >=SHAP (%)	97.83	98.42	98.38	98.31	97.76

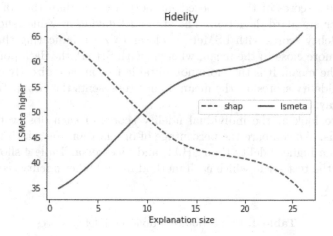

Fig. 4. Fidelity comparison (LSMeta vs SHAP)

In terms of fidelity, we observe that as the number of important variables to be returned increases, the fidelity of the proposed method is better than, if not equal to that of SHAP. In terms of stability, whatever the number of important variables to be returned, the stability of the proposed method is better than, if not equal to that of SHAP.

We think that it was somewhat predictable that the stability of the proposed method would be higher, because of the random factor in the selection of variable combinations to be used in the SHAP method. Indeed, with the proposed method there is a level of conservation of contributions which is maintained through the LRP method (The contributions of the weak learners do not change whatever

---

[3] In line, percentage of observations on which the proposed LSMeta method has a fidelity/stability score greater than or equal to SHAP. In columns, the percentage obtained for each number of explanatory variables to be returned by the methods.

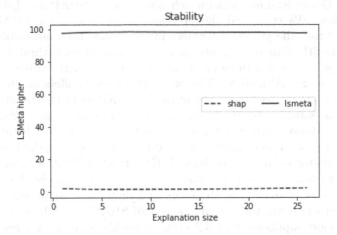

Fig. 5. Stability comparison (LSMeta vs SHAP)

the execution) which means that even if the SHAP method is used, its instability will be somewhat mitigated by the conservation of contributions provided
by LRP. This would justify that the proposed LSMeta method provides more
stable explanations than SHAP method. Even for fidelity, the random selection
of combinations of variables means that the combinations taken randomly to
calculate the contributions of variable i will not necessarily be the same combinations taken to calculate the contributions of variable j. If the combinations
chosen do not adequately express the weight of the features concerned by the
calculation, its contribution would be affected and instead of having a high/low
contribution, its contribution could be reduced/increased, which will have an
impact on the fidelity of the explanation. However, with the proposed method,
although the situation described above would also occur, the fixed contributions
of weak learners will mitigate this effect.

5   Conclusion

As companies emit the most of carbon dioxide through their various activities,
it becomes essential for them to reduce their amount of emissions. For this,
it is necessary to identify the factors that are responsible for it. This identification will be beneficial for a company: it would serve as a basis for defining
effective pollution reduction strategies, and for regulators: it would give indi-
cations on the aspects to be monitored within a company and even help to
distinguish high-risk companies/factors from those that are less so. Our work
consisted of predicting companies' CO_2 emissions and then identifying the main
factors that cause the amount of emission. For this we have built a Stacked
Generalization ensemble model made up of six weak learners and a meta-learner
to combine the predictions, then proposed an explainability method specific to

the Stacked Generalization model which combines the SHAP and LRP explainability method. We compared the proposed method with the SHAP method and obtained that the proposed LSMeta method provides more stable explanations than the SHAP method. It also provides explanations faithful to the model output and in some cases these explanations are even more faithful than those provided by the SHAP method. The proposed method suffers on the one hand from the same execution time problem as the SHAP method, and on the other hand from the same specificity to be applicable only to a single type of model as the LRP method. Furthermore, it is important to note that in this work we have used a restricted configurations to test the proposed method: we use one SHAP variant -KernelSHAP, the basic LRP variant -LRP-0, we test only on a regression problem, on a single dataset, we choose empirically the size of the explanation to be provided by the compared methods. Expanding the configurations used like using the specific variant of SHAP for each weak learner or using other more sophisticated LRP versions could constitute interesting directions of exploration. In regard to CO_2 emissions in general, it would be relevant to also consider the emissions generated by the training and usage of artificial intelligence models themselves [3].

Disclosure of Interests. We have no competing interests to declare that are relevant to the content of this article.

References

1. Adadi, A., Berrada, M.: Peeking inside the black-box: a survey on explainable artificial intelligence (XAI). IEEE Access **6**, 52138–52160 (2018)
2. Bach, S., Binder, A., Montavon, G., Klauschen, F., Müller, K.R., Samek, W.: On pixel-wise explanations for non-linear classifier decisions by layer-wise relevance propagation. PLoS ONE **10**(7), e0130140 (2015)
3. Delanoë, P., Tchuente, D., Colin, G.: Method and evaluations of the effective gain of artificial intelligence models for reducing co2 emissions. J. Environ. Manage. **331**, 117261 (2023)
4. Goldhammer, B., Busse, C., Busch, T.: Estimating corporate carbon footprints with externally available data. J. Ind. Ecol. **21**(5), 1165–1179 (2017)
5. Griffin, P.A., Lont, D.H., Sun, E.Y.: The relevance to investors of greenhouse gas emission disclosures. Contemp. Account. Res. **34**(2), 1265–1297 (2017)
6. Lundberg, S.M., Lee, S.I.: A unified approach to interpreting model predictions. Adv. Neural Inf. Process. Syst. **30** (2017)
7. Nguyen, D.: Comparing automatic and human evaluation of local explanations for text classification. In: Proceedings of the 2018 Conference of the North American Chapter of the Association for Computational Linguistics: Human Language Technologies, Volume 1 (Long Papers), pp. 1069–1078 (2018)
8. Nguyen, Q., Diaz-Rainey, I., Kuruppuarachchi, D.: Predicting corporate carbon footprints for climate finance risk analyses: a machine learning approach. Energy Econ. **95**, 105129 (2021)
9. Ribeiro, M.T., Singh, S., Guestrin, C.: "Why should i trust you?" Explaining the predictions of any classifier. In: Proceedings of the 22nd ACM SIGKDD International Conference on Knowledge Discovery and Data Mining, pp. 1135–1144 (2016)

10. Saleh, C., Chairdino Leuveano, R.A., Ab Rahman, M.N., Md Deros, B., Dzakiyullah, N.R.: Prediction of co2 emissions using an artificial neural network: The case of the sugar industry. Adv. Sci. Lett. **21**(10), 3079–3083 (2015)
11. Saleh, C., Dzakiyullah, N.R., Nugroho, J.B.: Carbon dioxide emission prediction using support vector machine. IOP Conf. Ser.: Mater. Sci. Eng. **114**, 012148 (2016). IOP Publishing
12. Van Buuren, S., Groothuis-Oudshoorn, K.: mice: Multivariate imputation by chained equations in r. J. Stat. Softw. **45**, 1–67 (2011)
13. Visani, G., Bagli, E., Chesani, F., Poluzzi, A., Capuzzo, D.: Statistical stability indices for lime: Obtaining reliable explanations for machine learning models. J. Oper. Res. Soc. **73**(1), 91–101 (2022)
14. Xu, G., Schwarz, P., Yang, H.: Determining china's co2 emissions peak with a dynamic nonlinear artificial neural network approach and scenario analysis. Energy Policy **128**, 752–762 (2019)
15. Yang, S., Wang, Y., Ao, W., Bai, Y., Li, C.: Prediction and analysis of co2 emission in Chongqing for the protection of environment and public health. Int. J. Environ. Res. Public Health **15**(3), 530 (2018)

Building and Validating an Ontology for Public Procurement: The Case of Cameroon

Jules Kouamo[1]([✉]) [iD], Etienne Kouokam[1] [iD], and Ghislain Atemezing[2] [iD]

[1] Computer Science Department, University of Yaounde I, Yaounde, Cameroon
{quentin.kouamo,etienne.kouokam}@facsciences-uy1.cm
[2] Mondeca, 75009 Paris, France
ghislain.atemezing@mondeca.com
https://facsciences.uy1.cm/ , https://mondeca.com

Abstract. Public procurement or tendering generally refers to the process followed by public authorities for the purchase of goods and services. In this process, the publication of tender notices online has become an essential and mandatory step in many countries, including in Cameroon. However, this step produces a huge amount of data, which makes the search for information tedious. Several solutions have been explored to overcome this problem, including the use of the semantic web, which allows web resources to be understood by machines. This paper presents the `APCO` ontology, a public procurement ontology that semantically describes the publication stage of the public procurement process. We use the `APCO` ontology to generate a knowledge graph for Cameroonian public procurement that will facilitate information retrieval and data analysis in Cameroonian public procurement.

Keywords: public market · tender · semantic web · ontology · open government data

1 Introduction

A public contract, whether awarded to a public entity or a private economic operator, meets the needs for works, supplies or services of the purchaser, whether it is a public entity (State, local authority, hospital, etc.) or a private entity. Public procurement legislation governs the rules that apply to these contracts, and in Cameroon, it is Decree 2018/366 of 20 June 2018 on the Public Procurement Code[1] which sets out the provisions.

The publication of public contracts is an essential and now mandatory step in many countries, including in Africa, where it has become digitised with the advent of the web. In Cameroon, for example, Decree N°2018/0001/PM of 05 January 2018 authorises the creation of a specific e-procurement platform for public contracts, named *"Cameroon On Line E-procurement System"*

---

[1] https://tinyurl.com/5ae29rt4, last consultation: July 2023.

P. Melatagia Yonta et al. (Eds.): CRI 2023, CCIS 2085, pp. 106–116, 2024.
https://doi.org/10.1007/978-3-031-63110-8_9

(COLEPS)[2]. This transition to digital technology has considerably increased the amount of data available, with around 11,600 documents relating to tender notices published each year.

This overabundance of online data poses a challenge in terms of finding relevant information. Traditional searches based on keywords are no longer sufficient to sort and interpret these massive volumes of data. This is where the concept of the Semantic Web, introduced by Tim Berners-Lee[3] in the 90's, comes in as a solution.

This article seeks to leverage the Semantic Web's capacity to make web resources understandable for both humans and machines, enhancing the effective structuring and interpretation of data. The specific focus is on applying sound semantic web practices, particularly in the context of Cameroon, to harness the substantial data available in public contracts. The aim is to facilitate the retrieval and analysis of information related to public procurement in Cameroon.

The article is organised as follows: the second section presents the public procurement domain, the third gives a state of the art of existing ontologies for public procurement, the fourth describes the development stages of the APCO ontology.[4] The fifth presents the validation procedures and principles, as well as the knowledge graph generated from the ontology. Finally, we conclude our work by making some recommendations.

The APCO ontology is freely searchable and accessible online.[5].

2 Public Procurement

According to the World Bank[6], public contracts are commercial transactions between governments and companies, in which governments buy goods, services or construction work from companies. It is a process that takes place in several stages.

2.1 Public Procurement Process

The public procurement process comprises several phases:

Preparation Phase: A public authority prepares the information and documents for the envisaged public procurement and a call for tenders. The public authority can leverage available data sources to search for similar contracts. This is useful when the authority is preparing the contract. For instance, it can obtain information on the usual prices of similar markets, constraints often associated with this type of market, etc.

---

[2] https://www.marchespublics.cm/index.do, last consultation: June 2023.
[3] https://www.w3.org/DesignIssues/LinkedData.html, last consultation: January 2023.
[4] https://ai4africa.github.io/apcoKG/documentation/index-en.html.
[5] https://ai4africa.github.io/apcoKG/, last consultation: January 2024.
[6] https://tinyurl.com/worldbanks.

Publication Phase: During this phase, the call for tenders is published on various platforms. In Cameroon, tender notices are published in the Public Procurement Journal, on the websites of the Ministry of Public Procurement (MIN-MAP)[7], and that of the Public Procurement Regulatory Agency (ARMP)[8]. The authority can also search for suitable suppliers in available data sources and invite them to submit their bids based on the type of tender. For example, searching for providers who have executed similar contracts with good evaluations. The authority then receives bids within a specified period and selects the best offer. Once the best offer is chosen, a contract is signed between the authority and the provider, and an award decision is published.

Evaluation Phase: Here, the public authority assesses the execution of the contract in relation to all given criteria. Additionally, the final and actual price of the contract is then known. The last two phases can be iterated until the contracting authority is satisfied.

The public procurement process involves a number of key players to ensure transparency and efficiency: the contracting authority, the procurement department, the tender committee, the candidates, the successful tenderer, the contracting authority, the prime contractor and the auditors. These players may vary according to the procedures and laws in force in each country.

The framework of our work being that of the semantic web, we are focusing on the publication phase of calls for tender in the public procurement process.

2.2   Limitations in the Public Procurement Process: The Case of Cameroon

The publication phase of the public procurement process in Cameroon faces a number of difficulties [9], including:

Difficulty in Finding Relevant Information: With a large number of published tender notices, it can be challenging for bidders to identify those pertinent to their activities. Moreover, tender publication notices are often distributed across different sites, making information retrieval difficult. In Cameroon, tender notices are published on two different platforms: the public procurement journal's website and the website of the Ministry of Public Procurement of Cameroon;

Difficulty in Understanding the Terms Used: technical or legal terms in tender notices can be challenging for bidders unfamiliar with the domain to comprehend;

Non-compliance of Publication Notices: Notices may be poorly formulated, lack necessary information, or contain errors;

Non-compliance of Submission Documents: Submission documents may be incomplete or fail to meet the requirements of the tender;

---

[7] https://minmap.cm/, last consultation: September 2023.
[8] https://www.armp.cm/, last consultation: October 2023.

Difficulty in Understanding Technical Specifications: technical specifications can be challenging for candidates to understand, leading to inappropriate bids.

To address these shortcomings, various solutions have been proposed, including the ontology development. Ontologies helps overcome these challenges by enabling better structuring of data in tender notices. By using a specific ontology, public entities can clearly describe their needs and employ standardised terms. This allows bidders to better comprehend the tender requirements and provide more precise and tailored bids. Ontologies can also facilitate the search for relevant tender notices, enabling bidders to navigate easily through different market categories and quickly find notices corresponding to their activities.

3   Ontologies and Public Procurement

An ontology is defined as a formal specification of a shared conceptualisation [1]. Ontologies help people and machines to communicate in a precise way to support the exchange of semantics. There are several in the field of public procurement.

PPROC (*Public Procurement Ontology*) [5] was created to contribute to the development of standards that can be used to publish all information deemed appropriate from a transparency perspective. It enables the provision of information recorded in the buyer profile of public entities and provides all relevant information regarding public procurement procedures.

The enumeration of terms relevant to the model was conducted with two sets of stakeholders: firstly, the company IASOFT, which developed buyer profiles for numerous Spanish administrations, and secondly, legal experts. Subsequently, the PPROC development team used these terms as a basis to prepare an initial list of entities, including cardinalities, domain, and range of properties.

The strategy employed for ontology development is BOTTOM-UP, and its implementation is done in OWL, following the method proposed by Noy et al. [8]. PPROC uses a primary class (Contract) that serves as the entry point to connect with other classes. Several classes from the PCO ontology are reused, and it considers that a contract can be subdivided into lots, making each lot a separate market. A contract is subject to selection criteria, which can be objective (low price, delivery time, etc.) or subjective (reputation, determined by experts). To describe the organization (stakeholders), properties, and vocabulary of other ontologies were used, such as Schema.org, Friend Of A Friend (FOAF), or SKOS. To define the content whose location or specific place must be known (e.g., the office of the contracting authority), the PLACE class was used along with Schema.org properties. To define the type of procedure and its urgency, SKOS was used.

In PPROC, the duration of a contract does not end with execution, meaning the moment when the procurement process is considered finished. Contracts are often subsequently modified through specific procedures, which often change aspects such as price or execution time. A block called ACCOMPLISHMENT has been created for this phase, due to the fact that regulations on contract publication vary by country, but for the sake of transparency, modifications are published.

LOTED2 [3] is a legal ontology that supports the identification of legal concepts and, more generally, legal reasoning. It was developed to address the question: 'How to design a legal ontology that meets the dual demand for improved access to legal information while preserving the richness of legal content, constraints, and peculiarities of the legal domain?' LOTE2 represents a compromise between a precise representation of legal knowledge and requirements for the web of data. Its source of information is the website that publishes all European public procurement notices, TED, which gathers approximately 1500 notices per week. The ultimate goal of the ontology is to enable the construction of semantic legal web applications that support public procurement by matching supply and demand. Two main directives cover the domain of European public procurement: Directive 2004/18/EC and Directive 2004/17/EC. The first regulates the coordination of public procurement procedures for works, supplies, and public services by contracting authorities (i.e., authorities operating in the so-called 'ordinary' sectors); the second regulates procurement procedures for entities operating in the water, energy, transport, and postal services sectors (i.e., the 'public services' sectors). These two legal sources serve as the reference point for deriving the exact meaning of terms used to describe the domain of public procurement and for extracting the information necessary to construct any logical theory formalizing domain knowledge. The design of LOTE2 is based on both a bottom-up approach (extracting legal concepts from legal sources) and a top-down approach (analyzing standard forms).

PCO ontology (*Public Contract Ontology*) [6] was developed with the aim of demonstrating the application of linked data for publishing contract-related data in the public sector, emulating the market process of matching supply and demand to generate 'business impact.' It encompasses information related to public contracts published by contracting authorities during the procurement process, based on the analysis of existing public procurement portals and the information published by contracting authorities on these portals. Not all information is considered, but primarily those that are useful for connecting public contracts and potential suppliers. Therefore, a significant portion of the information produced during the tendering phase (description of the public contract, received bids, and the ultimately accepted bid) is considered. Information resulting from other phases is only partially considered.

OCDS (*Open Contracting Data Standard*) [10] enables the publication of data and documents at all stages of the public procurement process by defining a common data model for structured publication. It was created to help organizations enhance contract transparency and enable in-depth analysis of contract data by a broad range of users. It is document-oriented and focuses on packaging and delivering relevant data iteratively and eventfully through a series of versions.

Table 1. Summary of ontologies for public procurement.

Reviewed ontologies	PPROC	PCO	OCDS	LOTED 2
required granularity	✓	X	✓	X
Complexity	X	✓	X	X
Objective	X	✓	✓	X
Context	X	X	X	X
Number of classes-properties	78–130	22–51	24–142	22–101

Table 1 present a comparison of four popular ontologies for tender on several criteria according to [8]. None of the ontologies studied is adaptable to an African context because the resources used to develop them are exclusively European, for example approximately 20% of the concepts manipulated by PPROC come from Spanish law, and the others from the TED public procurement platform. Which means that most of the concepts manipulated in these are not semantically close to African concepts. As an illustration, we have the concept of 'Project Owner' which, although similar in the way it is implemented, does not play the same role in both contexts.. To improve the structuring of Cameroonian public procurement data, it is therefore necessary to develop a specific ontology.

4   APCO: An Ontology for African Public Procurement

To build our domain ontology, the methodology developed by [7] has been used. It is a incremental methodology made up by 7 steps:

1. Define the domain and scope of the ontology;
2. Consider the potential reuse of existing ontologies;
3. List the important terms in the ontology;
4. Define the classes and the hierarchy of classes;
5. Define the properties of classes-attributes;
6. Define the facets of attributes;
7. Create the instances

The first step is to define the scope and domain of the ontology and therefore answer the following questions:

- **Q1** what domain will the ontology cover?
- **Q2** for what purpose will we use the ontology?
- **Q3** What types of questions will the ontology have to answer?

(Q1): The domain covered by the ontology is that of public procurement, aiming to provide consensual knowledge and an information model related to public procurement, specifically those in Cameroon.

(Q2): Starting from the observation that research is still conducted in a conventional manner on various publication sites for documents such as tender notices and other public procurement-related documents, which hinders the production of relevant information. The ontology we are developing aims to search for pertinent information based on metadata describing documents related to public procurement.

(Q3): To validate the process, competency questions[9] were formulated and validated with the help of an expert in the field of public procurement. "Subsequently, we identify ontological and non-ontological resources. As non-ontological resources, we have Decree No. 2018/366 of June 20, 2018, establishing the Public Procurement Code1, which governs public procurement in Cameroon. It contains the vocabulary used for today's public procurement. To date, there is no RDF version of this vocabulary. We can also mention Order No. 004/CAB/PM of December 30, 2005, concerning the implementation of the Public Procurement Code. Finally, ontological resources include, among others, PCO, PPROC, Good Relation, Dublin Core, Geofla and DBpedia.

Following the definition of the domain and scope of the ontology, the next step is to list the terms and concepts of the ontology. We use documents governing public procurement in the Central African sub-region, as well as data from the websites of various public procurement regulatory agencies, to list the terms and concepts that should be used in the ontology. This step was carried out by ontology developers and some stakeholders in public procurement. The output of this step is a list of terms that will be used to successfully proceed to the next step in the methodology, namely the possible reuse of ontologies.

Indeed, with the recent development of the semantic web, there is a multitude of vocabularies available online, along with several web applications that catalogue these different vocabularies. This is the case with LOV [11], which gathers millions of vocabularies available on the internet. We use this platform to explore potential ontologies that we will reuse for the development of our ontology. It is sufficient to enter a term into the platform and see if the term already exists in an ontology, and then check if its description aligns with the semantics desired in the ontology. For example, the term "VILLE" (CITY), which belongs to our initial list from the phase of establishing a list of terms. This term had already been described in a vocabulary (DBpedia), and its type and description had the same semantics that we wanted to highlight in the ontology. Therefore, the DBpedia ontology was reused. Table 2 presents all the concepts reused in various other ontologies.".

---

[9] https://github.com/AI4Africa/apcoKG/blob/main/spec/CompetencyQuestions. md, last consultation: November 2023.

Table 2. Reused concepts and properties

Ontologies	Concepts
DBpedia	City, Department, Region, department, locationCity
FOAF	Person
GoodRealtion	BusinessEntity, Offering
Public Pontrat(PC)	bidder, contractactingAuthority, procedureType
geofla	region
dcterms	date, identifier, title
PPROC	numberOfLots

We then define the classes, their hierarchy and properties. The main class is `apco:Publication`, representing documents related to public procurement. We use a "TOP-DOWN" approach to make hierarchy of classes, starting with the most general ones and specifying their characteristics. Properties, such as `apco:contractingAuthority` linking a tender notice to the contracting authority, are also defined. The facets of the attributes and properties are established, such as the cardinality 0..1 of `apco:makesOffer` meaning that a bidder can make at most one offer. Figure 1 shows an overview of part of the ontology.

The ontology is available online and can be consulted and used in the Linked Open Vocabulary (LOV).[10] [11].

5   Ontology Validation and Knowledge Graph

5.1   Ontology Validation

APCO has been developed following the ontology construction rules and principles defined by [4]: Completeness is ensured by definitions associated with concepts and relations, using necessary conditions, or even necessary and sufficient conditions where possible. In addition, ontological extensibility is maximised, ensuring that each term has a specific definition that distinguishes it from more general or specialised terms. This enables a robust and accurate ontology to be established, promoting a comprehensive representation of the Cameroon public procurement domain. To validate the structure [2] of our ontology, we are using two tools, *OntoCheck* and *Pellet*. These extension modules for the PROTEGE ontology editor are designed to check compliance with naming conventions, metadata completeness, cardinalities, labels and ontology metrics.

Unlike the ontologies studied in Table 1, our approach for the APCO ontology involved the use of resources specific to the Cameroonian and African context. This approach favours a more appropriate interpretation of the concepts

---

[10] https://lov.linkeddata.es/dataset/lov/vocabs/apco, last consultation: December 2023.

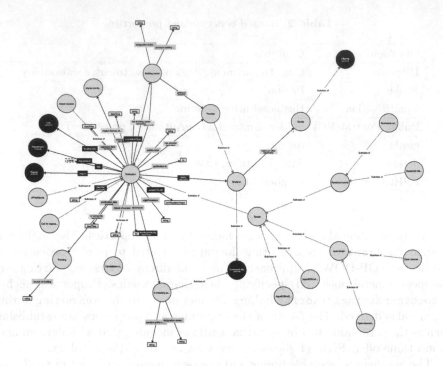

Fig. 1. Ontology visualisation

and relationships within the ontology by public procurement stakeholders in Cameroon.

The ontology was designed by answering all the competence questions in the form of SPARQL queries[11], demonstrating its effectiveness. In addition, its design has been guided by the FAIR principles, guaranteeing that the data is accessible, interoperable, reusable and easy to share.

5.2   Knowledge Graph

The aim is to use the ontology developed to generate a knowledge graph of public procurement in Cameroon. The experimental data used comes from ARMP and includes contract notices, award decisions and other publications on public procurement[12] of Cameroon for the years 2016 to 2021. It consists of 57589 rows and 20 columns. The knowledge graph is generated in several phases, as shown in the Fig. 2.

Data Acquisition: During the data collection phase, we utilize information from the public procurement database as well as certain data retrieved from

---

[11] https://github.com/AI4Africa/apcoKG/blob/main/spec/CQs.rq, last consultation: December 2023.
[12] https://www.armp.cm/.

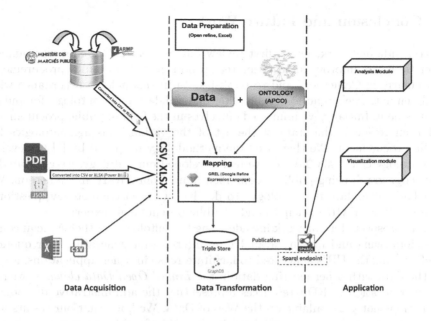

Fig. 2. Knowledge graph generation architecture

files, which are then transformed into the desired format using the PowerBI tool.

Data Transformation: The transformation of this data by formatting it according to specific needs, in particular using the OpenRefine tool to define data elements such as URIs and store them in a triple store (Ontolinga's GraphDB) growing from 57,589 rows at the outset to 778,688 corresponding to our triplets.

Application: In the application phase we use the generated knowledge graph[13] in the form of triples to develop smart applications such as to improve searching information on public contracts, and visualise and analyse data.

The knowledge graph generated in this way will provide a structured representation of data relating to contracts, suppliers, calls for tender and so on. This makes it easier to navigate between the different entities and their relationships, helping public procurement players to discover related information more efficiently. In addition, thanks to the semantic links between concepts, the knowledge graph helps to resolve ambiguities and refine searches, improving the relevance of results.

---

[13] https://linkedvocabs.org/dataset/dump_apcoKG.ttl, last consultation: April 2024.

6   Conclusion and Future Work

To conclude our work, let us first review our approach. Our aim was, among other things, to improve information retrieval in the field of public procurement, particularly in Cameroon, and to publish data in this field in accordance with good semantic web practice by building and validating an ontology for public procurement. Initially, we immersed ourselves in the field of public procurement, and then reviewed the state of the art of the various existing ontologies for public procurement. We then used the methodology proposed by [7] to develop our ontology called APCO. Once we had developed our ontology, we generated a knowledge graph using public procurement data from ARMP in Cameroon. We then had to validate the ontology. To do this, we used competency questions, inference tools, FAIR principles and a public procurement expert.

The prospects for our work include using the ontology in other African countries, improving and making available an interface for generating user queries, dereferencing the URIs generated to facilitate reuse in other applications, aligning the data with other existing data in the *Linked Open Data cloud*, as well as the incorporation of RDF resources representing the administrative division of our regions not yet available on the Web of Data. We hope that our results will help to improve information retrieval in the field of public procurement in the future.

References

1. Borst, W.N.: Construction of engineering ontologies for Knowledge sharing and reuse. Ph.D. thesis, University (1997)
2. Despres, S.: OOGO: ontologie des outils utiles à la gestion d'ontologies. In: 29es Journées Francophones d'Ingénierie des Connaissances, IC 2018 (2018)
3. Distinto, I., d'Aquin, M., Motta, E.: Loted2: SN ontology of European public procurement notices. Semantic Web **7** (2016)
4. Gruber, T.R.: A translation approach to portable ontology specifications. Knowl. Acquisit. (1993)
5. Muñoz-Soro, et al.: PPROC, an ontology for transparency in public procurement. Semantic Web **7** (2016)
6. Nečaskỳ, M., Klímek, et al.: Linked data support for filing public contracts. Comput. Indust. **65** (2014)
7. Noy, N., McGuinness, D.L., et al.: Ontology Development 101. Stanford University, Knowledge Systems Laboratory (2001)
8. Noy, N.F.: Semantic integration: a survey of ontology-based approaches. Comput. Indust. **33** (2004)
9. Ntembe, A., et al.: Analysis of public investments and economic growth in cameroon. J. Econ. Financ. **42** (2018)
10. Soylu, A., et al.: Towards an ontology for public procurement based on the open contracting data standard. In: Conference on e-Business, e-Services and e-Society (2019)
11. Vandenbussche, P.Y., Atemezing, G.A., et al.: Linked open vocabularies (LOV): a gateway to reusable semantic vocabularies on the web. Semantic Web **8**(13) (2017)

A Hybrid Algorithm Based on Tabu Search and K-Means for Solving the Traveling Salesman Problem

Leopold Kamchoum Nkwengoua[ID] and Mathurin Soh[(✉)][ID]

University of Dschang, Dschang, Cameroun
mathurinsoh@gmail.com, kamchoumleopold@gmail.com

Abstract. In this paper, we propose an approach to solve the symmetric Traveling Salesman Problem (TSP) by combining the K-means clustering technique and tabu search (TS). In this hybrid approach, we first apply the K-means algorithm to group cities into several clusters. Then we use tabu search to explore the solution space to optimise the path within each cluster. This avoids getting stuck in local optima and allows us to explore new, potentially better solutions. Finally, to combine the solutions from the different clusters, we apply a recombination technique that randomly selects a cluster from the list of clusters, then calculates the distance between one end of this cluster and the ends of the other clusters. It then chooses the smallest distance between the previously calculated distances, and recombines the two clusters to obtain a new cluster. The operation is repeated until all the clusters have been reunited. The experiments, carried out on instances taken from the TSPLIB library [15, 17], show that the hybrid approach proposed in this way provides a significant improvement in terms of the length of the path travelled and the travel time, compared with methods based solely on K-means or tabu search.

Keywords: Clustering · Hybrid · K-means · Recombination · Tabu search · TSP

1 Introduction

The electronic circuit design industry continues to revolutionize the way we live, with tools such as cell phones, computers, autonomous cars and many others that are becoming ever smaller. However, this science faces a major problem in electronic circuit design. This problem is optimizing the order in which components are placed on an electronic chip. In other words, we need to determine the ideal way to place electronic components on a chip. This would considerably improve circuit performance, by limiting electromagnetic interference between components, and reducing power consumption. This problem is basically identical to the Travelling Salesman Problem (TSP), according to Liu Tian-shu et al. [12]. There are other problems that are also basically identical to the TSP, such as

P. Melatagia Yonta et al. (Eds.): CRI 2023, CCIS 2085, pp. 117–128, 2024.
https://doi.org/10.1007/978-3-031-63110-8_10

vehicle route planning, fleet management, parcel distribution, waste collection, and soon.

TSP is an NP-complete problem [5], which means that there is no algorithm capable of determining the exact solution in a reasonable time. Many researchers have developed solution methods such as tabu search [1], simulated annealing [7], the ant colony algorithm (ACO) [6], genetic algorithms [8], etc., to solve this problem. Some of these methods are very effective on small instances of the TSP, others on medium-sized instances, and others on large instances. We propose to develop a hybrid approach that is highly efficient in terms of solution quality and time on both medium and large TSP instances.

The remainder of this article is organized as follows. In Sect. 1, we present the problem of solving the TSP and its application to printed circuit board design. In Sect. 2, we present the TSP, its mathematical formulation, and some applications of this problem. In Sect. 3, we present methods for solving the TSP that exist in the literature. In Sect. 4, we present the algorithms we use in this paper to solve the TSP. In Sect. 5, we present our hybridization technique and an illustration of how it works. In Sect. 6, we present the experimentation of our solution, the results, and the analysis of the results obtained. Finally, we conclude with a review of the work carried out, its limitations, and prospects for future improvements.

2   Problem Statement

TSP was first presented as a game by William Rowan Hamilton in 1859. In its most classic form, it reads as follows: "A commercial traveler must visit a finite number of cities once and return to his point of origin. Find the order in which the cities are visited that minimizes the total distance traveled by the traveler". Formally, the TSP can be defined as a complete graph $G = (V, E)$ where V is the set of nodes (cities) in the graph and E is the set of edges in the graph. The graph is made up of a series of edges representing the distance between two nodes or cities. In terms of graph theory, solving the TSP means searching for a cycle or Hamiltonian tour of minimum length. This can be done by enumerating all the cycles in the graph and their respective lengths, and then selecting the cycle with the smallest length. This method gives us the best results in a reasonable time for a very limited number of cities, ranging from 10 to 20. On the other hand, for a relatively large number of cities, it takes a considerable time to obtain the best results. This is because, knowing that N represents the number of cities in the TSP instance, once the starting city has been determined, (N-1) remains! Ways to determine the second city, and (N-2)! Ways to determine the third city, and so on. If each solution found is symmetrical (case of symmetrical TSP), we have (N-1)!/2 possible solutions. This means that the complexity of this exact solution method is N!

2.1   Matematical Formulation

According to [9], the TSP can be defined on a complete undirected graph G = (V, E) with n vertices, where each vertex represents a city and each edge (i, j) has a weight w(i, j) representing the distance between cities i and j. Let x(i, j) be a binary variable which takes the value 1 if edge (i, j) is included in the Hamiltonian cycle and 0 otherwise. Let u(i) be a continuous variable representing the position of city i in the Hamiltonian cycle. The objective function is then:

min

$$\sum_{(i,j)\in E} w(i,j)x(i,j)$$

under the constraints :

$$\sum_{(i,j)\in E} x(i,j) = n$$

$$\sum_{j\in E} x(i,j) = 2$$

$$\forall\, i \in V$$

$$u(i) - u(j) + nx(i,j) \le n - 1 \,\forall\, (i,j) \in E, i \ne 1, j \ne 1$$

$$1 \le u(i) \le n \,\forall\, i \in V$$

where the first constraint guarantees that the Hamiltonian cycle passes through all cities, the second constraint guarantees that each city is visited exactly once, the third constraint prevents the formation of sub-cycles, and the fourth constraint defines the order of cities in the Hamiltonian cycle.

2.2   Application of TSP

The Traveling Salesman Problem (TSP) has applications in a variety of fields [10,12], notably in logistics and transport, where it is used to optimize delivery routes, vehicle tour planning, etc. Printed circuit board routing, where it is used to determine the order of placement of components on an electronic chip. In tour planning, where it is used in the tourism industry to plan efficient tours that visit multiple attractions while minimizing distances traveled. Also in drone route planning, where it is used to optimize drone flight paths for surveillance, delivery, mapping and other missions.

3   State of the Art

TSP is an NP-complete problem [5], as there is no algorithm capable of finding an exact solution in a reasonable time. These scientists have carried out several works aimed at developing methods that provide approximate solutions in a reasonable time. Among these works Glover, F. [1], who in 1989 developed a meta-heuristic called Tabu Search TS. This method operates according to an exploration phase of the solution space, where it uses local search [11] to determine candidate solutions, and an exploitation phase, where it chooses the best candidate solution from the existing candidate solutions. This method is very efficient in terms of time and solution quality on small TSP instances of up to 100 cities. But it is less time-efficient on medium and large TSP instances. This is due to the fact that during the exploration phase, the algorithm must enumerate all candidate solutions according to the formula $(N(N\text{ -}1))\ /2$ using local search. N represents the number of cities, so the greater the number of cities, the greater the number of candidate solutions. This makes the program increasingly inefficient. For this reason, we plan to use a technique like k-means to reduce the number of cities during the exploration phase.

Dewantoro et al. [2] propose in 2019 a combination of the ant colony algorithm (ACO) and tabu search (TS), to solve the TSP. The authors exploit the power of the ACO to efficiently explore the solution space using multiple ant colonies, which provides a good initial solution for the tabu search algorithm, they also exploit the power of the TS to improve already existing solutions. This gives the algorithm a power that surpasses ACO and TS performed individually in terms of time and solution quality, as shown by the results obtained in [2]. However, these results also reveal that this hybrid algorithm is less efficient on medium and large TSP problems. This can be explained by the fact that this algorithm does not solve the TS exploration phase flaw. As we said in [1], this makes the algorithm less efficient as the number of cities becomes too large.

Wu et al. [3] propose in 2019 to use deep reinforcement learning to solve the TSP. Reinforcement learning is used by the authors to determine a policy that is capable at each step, of choosing the next best city to visit, in order to obtain the best possible route at the end of its execution. This algorithm takes city coordinates as input. Then choose the starting city at random. Then the algorithm uses the negative length of the tour as a reward signal. This algorithm performs very well in terms of time and solution quality on small TSP instances of up to 50 cities. On the other hand, it performs less well on medium and large TSP instances in terms of time. This is because the training phase needs to take into account the calculation of value functions and optimal policies, which become more numerous as the number of cities increases. This greatly increases the time taken by the algorithm to obtain a good-quality solution.

Diabat, W. [4] propose in 2019 the Firefly algorithm (FA) and K-means clustering to solve TSP. This algorithm proposed by the authors works in three stages. The first step is to group the set of cities into a finite number of clusters, the second step is to iteratively apply the FA algorithm in each cluster, in order to obtain solutions in each cluster called optimal sub-solutions. Then, in the third step, the set of optimal sub-solutions is combined into a single global optimal solution. The results obtained by the authors in [4] show that the algorithm performs very well in terms of time and solution quality on medium and large TSP instances. Nevertheless, these results could be further improved by conducting in-depth studies to determine the ideal parameters for the algorithm.

4   Presentation of K-Means and Tabu Search

4.1   Presentation of Tabu Search

Candidate Solutions Exploration Phase. The exploration phase consists in enumerating all candidate solutions from an initial solution, using the 2 Opt local search algorithm, which consists in changing the position of two randomly chosen cities in the initial solution. This produces new candidate solutions on each iteration according to the formula $(N(N-1))/2$, where N represents the number of cities in the TSP instance in question. The image below illustrates how the exploration phase works on a single iteration (Fig. 1).

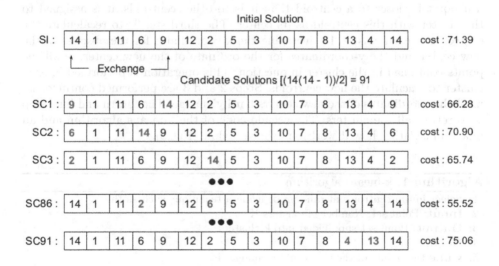

Fig. 1. Running the TS algorithm over one iteration.

Exploitation Phase for Candidate Solutions. In the exploitation phase, we select the solution with the lowest cost among the candidate solutions listed during the exploration phase. In the present case, the SC86 solution is the one with a lower cost than the other candidate solutions proposed, and given that it is non-taboo, i.e. not contained in our list of taboos (a contrario, we would choose the solution directly superior to SC86 and which is non-taboo, which means that we very often obtain a solution that is of lower quality than the previous solution, ensuring that we don't get stuck in local optima). Since our list is currently empty, it is the best solution for this first iteration. Consequently, it will be the initial solution for the second iteration and will be saved in our tabu list. The algorithm will perform the same operations until the number of iterations has been reached. It will then return the best solution and its cost.

4.2 Presentation of the Clustering Algorithm (K-Means)

K-means is an algorithm particularly used in machine learning, especially in unsupervised learning, to separate a finite data set into finite data groups according to a certain criterion such as the distance between the different data. We use it in our approach to exploit this property, because, as we said in [1], reducing the number of cities during the exploration phase will undoubtedly enable MT to obtain good results in a shorter time.

This algorithm works in four phases. The first phase consists of choosing the centroids of our various clusters at random, or using a technique such as eblow's bend [14]. The second step consists in forming the different clusters by assigning to each cluster the point that is closest to its centroid. In other words, if a point is closer to a centroid than it is to other centroids, it is assigned to the cluster with this centroid as its center. The third step is to recalculate the centroids of each cluster, by averaging the x-coordinates for the abscissa of the new center and the y-coordinates for the ordinate of the new center, of all the points contained in the cluster in question. This operation is performed in each cluster to calculate the new centroids. Steps 2 and 3 are performed continuously until the coordinates of the new centers hardly change. The fourth and final step is to return all the clusters. The pseudo-code of the k-means algorithm and an image illustrating how it works are shown below (Fig. 2).

Algorithm 1. k-means algorithm

1: Select initial centroids at random or using a technique such as the
2: **Input:** Point set, Number of clusters K
3: **Output:** Point set subdivided into k clusters
4: **Begin**
5: **while** the points of the new centers change **do**
6:     **for** every point **do**
7:         Assign the point to the cluster with the nearest centroid
8:     **end for**
9:     **for** every cluster **do**
10:        Recalculate the centroid points by averaging the coordinates of the cluster points
11:    **end for**
12: **end while**
13: Return all clusters
14: **End**

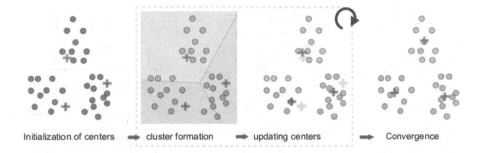

Initialization of centers ➡ cluster formation ➡ updating centers ➡ Convergence

Fig. 2. Illustration of the K-means steps, source [13].

5 Proposal of a Hybrid Approach Combining Tabu Search and K-Means for Tsp Resolution

5.1 Operating Principle of Our Approach

The operating principle of our algorithm is as follows: first, the K-means is initially used to group the cities of the TSP instance into K clusters. K here represents the number of clusters, and is initially set according to the number of cities in the TSP instance. Each cluster represents a group of cities that will later be visited by the TS. The TS takes a number of parameters, such as the initial solution or initial tour. We generate this initial tour by randomly shuffling the list of cities to be visited. Once the initial tours have been generated, the TS is used to improve the quality of solutions within each cluster. To do this, it successively explores each cluster and makes local moves (generating a local candidate solution), then uses a tabu list to avoid solutions already explored. Once the TS has completed the cluster exploration, during which it has determined the best solutions for each cluster, a recombination algorithm is applied to recombine the clusters into a single one so that the final route is as optimal as possible. The operating principle of this algorithm is inspired by that of the nearest neighbor algorithm. The recombination algorithm stochastically selects one cluster from the others, then stochastically selects one end (city) of the selected cluster, given that each cluster has two ends. Then, starting from this endpoint, it calculates all the possible distances between this endpoint and all the other endpoints of the other clusters. Once the calculation is complete, the program chooses the smallest of all the distances calculated. This distance is the smallest distance between the two clusters. The two clusters are linked and become a single cluster. In the event that not all clusters are connected, the operation starts again, and stops only when all clusters are connected into a single cluster. The algorigram of this algorithm is as follows (Fig. 3).

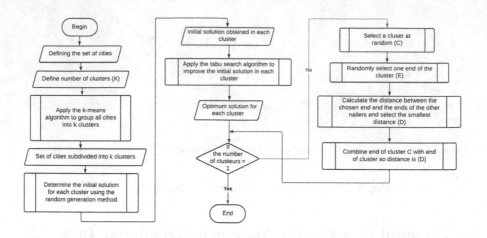

Fig. 3. Flowchart of our hybrid approach combining TS and K-means.

Algorithm 2. Kmeans-Tabu-Search algorithm

1: **Input:** Number of clusters K, number of iterations N , cities of the TSP instance
2: **Output:** Optimum final route
3: **Begin**
4: Use K-means to group cities into K groups
5: **for** each cluster **do**
6:     Generate an initial tour using the random generation technique
7:     Using tabu search to improve solution quality within the cluster cluster, with a number of iterations N
8: **end for**
9: **while** $|clusters| > 1$ **do**
10:     $cluster \leftarrow$ stochastically select a cluster from $clusters$.
11:     $extremite \leftarrow$ choose an end of $cluster$ stochastically.
12:     $distances \leftarrow$ calculate distances between $extremite$ and all other extremities of other clusters.
13:     $minDistance \leftarrow$ find the minimum distance among $distances$.
14:     connect the two clusters corresponding to $minDistance$.
15:     update $clusters$
16: **end while**
17: **End**

In our approach, we use High-level Relay Hybrid (HRH). High-level because we don't modify the internal operating principle of the tabu search and k-means algorithms, and relay because the k-means algorithm runs first and passes the baton to the tabu search algorithm.

5.2   Illustration of Operating Principle

We have just presented the operating principle of our approach in Sect. 5.1. In this section, we present a detailed illustration of the operating principle of our approach (Fig. 4).

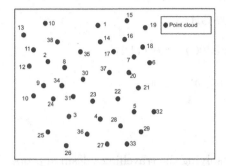

(a) Point cloud (city) of the TSP instance of 38 cities.

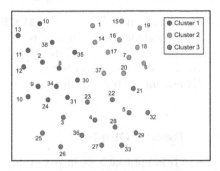

(b) Instance of 38 cities grouped into 3 clusters.

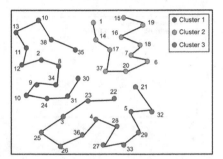

(c) Best route for each cluster determined by TS.

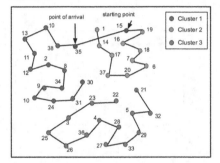

(d) Recombination of clusters 1 and 2 into a single cluster.

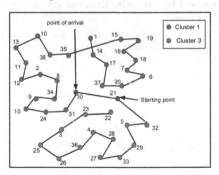

(e) Recombining the new cluster with another cluster

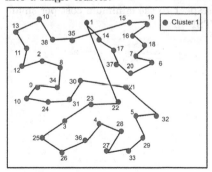

(f) Final itinerary obtained after recombining all clusters.

Fig. 4. Illustration of how our hybrid approach works.

6   Experimentation

6.1   Experimental Environment

To test our approach, we selected a few Euclidean instances from TSPLIB [15,17], whose best solutions are known and published in the literature. We performed the tests on a computer with the following characteristics:

- Processor: Intel Core i7-3540M CPU 3.00 GHz x 4
- RAM memory: 15.5 Giga byte
- Graphics card: NVC1/Intel HD Graphics 4000 (IVB GT2)
- Hard disk: 500.1 Giga byte
- Operating system: Ubuntu 20.04.1 LTS 64 bit

6.2   Results Obtained

The hybrid approach we have developed for solving the travelling salesman problem has enabled us to obtain the following results (Fig. 5).

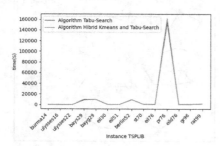

(a) Evolution of the Tabu-Search and Kmeans-Tabu-Search algorithms with respect to the minimum cost per small TSP instance.

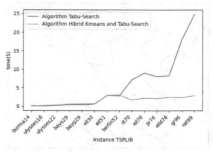

(b) Evolution of the Tabu-Search and Kmeans-Tabu-Search algorithms with respect to minimum time per small-size instance.

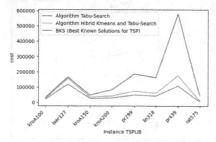

(c) Evolution of the Tabu-Search and Kmeans-Tabu-Search algorithms with respect to the minimum cost per instance of average size.

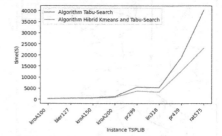

(d) Evolution of the Tabu-Search and Kmeans-Tabu-Search algorithms with respect to the minimum time per instance of average size

Fig. 5. Experimental results obtained with our hybrid approach.

6.3   Results Analysis

these results, which we have just presented in the figure above, were obtained after first running our hybrid algorithm for each TSP instance 10 times, and then tabulating the quality of the smallest solution and the corresponding execution time for each of the 10 results for each TSP instance, for each of the 10 results per TSP instance, we also tabulated the mean value of the solution quality and the mean value of the time taken to obtain these solutions, and these results (mean value) are shown in the figure above. These results show that on small TSP instances, our hybrid approach achieves, on the whole, the same results as the tabu search algorithm performed individually, but in a much better time than the tabu search algorithm as the problem size increases. Also, we note that on medium-sized instances of the TSP, our approach achieves results that are better than that of the individually executed tabu search algorithm. Although these results are not equal to the best solution currently known in the literature (BKS) [16,17], the time taken to obtain these solutions is still unknown in the literature. The time taken by our approach to obtain these results remains much better than that of tabu search.

7   Conclusion

In this article, we have presented our proposed approach to solving the TSP. We explained that the operating principle of this approach is based on three factors: the first factor consists in grouping the cities into several distinct city clusters; the second factor consists in applying the TS to each cluster in order to obtain optimal sub-solutions. and finally, the third factor consists in re-grouping all the clusters into a single one. We also presented the results we obtained, noting that our approach is much better than tabu search on large and medium instances in terms of time and cost. However, when we compare our results with the best results presented in the literature, we find that they are better than ours in terms of solution quality, but not necessarily in terms of time, as these results, published in the literature, are not accompanied by the respective execution time for each solution found. To improve the performance of our approach, we are considering.

- In the short term, we need to choose an ant colony algorithm to obtain the initial solution in each cluster, because up to now, we've always obtained it using the random generation method, and the quality of the results returned by the TS is highly dependent on the quality of the initial solution.
- In the long term, we could include parallelism in the second phase of our algorithm, which consists of applying the TS in each cluster iteratively. Including parallelism in this phase would allow us to apply the TS in each cluster, not iteratively, but in parallel this time. This would save us a considerable amount of time.
- In the long term, we need to carry out an in-depth study to find the right parameters for our approach. Such as the number of clusters to be chosen

for the different instances, and the number of iterations that TS will have to perform in order to obtain good results.

References

1. Glover, F.: Tabu search-part II. 2, 4–32 (1990). https://doi.org/10.1287/ijoc.2.1.4
2. Dewantoro, R.W., Sihombing, P., et al.: The combination of ant colony optimization (ACO) and tabu search (TS) algorithm to solve the traveling salesman problem (TSP), pp. 160–164 (2019). https://doi.org/10.1109/ELTICOM47379.2019.8943832
3. Wu, Y., Song, W., Cao, Z., Zhang, J., Lim, A.: Learning improvement heuristics for solving the travelling salesman problem (2019). https://doi.org/10.1109/TNNLS.2021.3068828
4. Jaradat, A., Diabat, W., et al.: Solving traveling salesman problem using firefly algorithm and k-means clustering, pp. 586–589 (2019). https://doi.org/10.1109/JEEIT.2019.8717463
5. Goubault-Larrecq, J.: Classes de complexité randomisées (2008)
6. Dorigo, M., Maniezzo, V., Colorni, A.: Ant system: optimization by a colony of cooperating agents. 26, 29–41 (1996). https://doi.org/10.1109/3477.484436
7. Kirkpatrick, S., Gelatt Jr, C.D., Vecchi, M.P.: Optimization by simulated annealing. 220, 671–680 (1983). https://doi.org/10.1126/science.220.4598.671
8. Geiger, M.J.: Genetic algorithms for multi-objective vehicle routing (2008). https://doi.org/10.48550/arXiv.0809.0416
9. Matai, R., Singh, S.P., Mittal, M.L.: Commercial traveller problem: an overview of applications, formulations and solution approaches. 1, 1–25 (2010)
10. Davendra, D.: The travelling salesman problem: theory and applications (2010)
11. Estellon, B., Gardi, F., Nouioua, K.: Vehicle scheduling: a large neighborhood local search approach, pp. 21–28 (2005)
12. Bousri, I., Boucetta, A., Sahabi-Abed, S.: Regionalization of annual temperature normals in Algeria using the K-means method 6, 40–46 (2022)
13. Reinelt, G.: TSPLIB-A traveling salesman problem library. 3, 376–384 (1991). https://doi.org/10.1287/ijoc.3.4.376
14. Lin, Y., Bian, Z., Liu, X.: Developing a dynamic neighborhood structure for an adaptive hybrid simulated annealing–tabu search algorithm to solve the symmetrical traveling salesman problem. 49, 937–952 (2016). https://doi.org/10.1016/j.asoc.2016.08.036
15. 1. Reinelt, G.: TSPLIB 95 documentation (1995)
16. Lin, S., Kernighan, B.W.: An effective heuristic algorithm for the traveling-salesman problem. 21, 498–516 (1973). https://doi.org/10.1287/opre.21.2.498
17. Tan, L.-Z., Tan, Y.-Y., Yun, G.-X., Zhang, C.: An improved genetic algorithm based on k-means clustering for solving traveling salesman problem, pp. 334–343 (2017). https://doi.org/10.1142/9789813200449_0042

Hybridization of a Recurrent Neural Network by Quadratic Programming for Combinatory Optimization: Case of Electricity Supply in a University Campus

Franck-steve Kamdem Kengne[✉][iD], Mathurin Soh[iD], and Pascaline Ndukum

University of Dschang, Dschang, Cameroon
franck.kamdem@univ-dschang.org

Abstract. Joule loss is a major concern in the power grid because it can lead to significant energy waste. There are a number of methods for minimizing joule loss, but they are still incomplete because they don't take into account the changing trend of electricity demand in buildings. In this paper, we propose a novel approach to minimizing joule loss using a hybridization of a recurrent neural network (RNN) and quadratic programming (QP). The RNN is used to learn the temporal dynamics of the electricity demand, while the QP is used to solve the combinatory optimization problem of minimizing joule loss. The recurrent neural network set up is able to predict the needs of the different campus premises in order to make the system flexible to the needs of consumers. It contains 5 layers: 3 of the Long short term memory type, 1 of the dropout type and 1 of the dense type for a total of 86,910 parameters to be determined during training. It was trained and validated using American hourly electricity power data normalized between 0 and 1. We model the power losses by Joule effect occurring during the transport from the production sources and to the consumers with quadratic programming. Once the optimization model was formulated based on electric laws, it was solved to determine the optimal current flows that should circulate in one part of the network. After simulation, the analysis of the results obtained shows that the proposed solution produces good results to the extent that by integrating variations in user needs, it makes it possible to determine the configuration which reduces hourly losses. To improve the accuracy of the proposed system one could use a data set containing a few more parameters affecting the consumption of the load such as meteorological datas.

Keywords: Combinatorial Optimization · Hybridization · Quadratic Programming · Recurrent Neural Network

https://www.univ-dschang.org/.

1 Introduction

Currently, electrical energy is one of the most demanded and used resources in the world as long as the global energy demand continues to increase (about 2% per year [1]). It is then necessary to ensure its availability at all times. However, in a power grid, the transmission of electricity from sources of production towards the users, there are losses of power under form of heat: this is the Joule effect. One of the main research challenges faced in electricity supply aims at providing electricity to consumers according to their usage while minimizing as much as possible power loss. Many approaches have been proposed to solve this problem by different mathematical methods and by the use of certain simplifications and special treatments: the Gradient method, quadratic programming, Newton's method and the interior point technique [2]. However, all these techniques have a lot of problems such as: convergence towards a local solution, large number of iterations, sensitivity to an initial search point, limited modeling capacity (in the management of constraints, hourly variation of electric power needs and functions not linear and discontinuous). Moreover, these different approaches do not take into account variations in consumer needs as long as at different times of the day, a user does not necessarily have the same needs. So, in the process of distributing electrical energy, how to minimize power losses due to Joule effect while taking into account variations in user needs? By considering a campus made up of several buildings and sources of electrical energy production, we aim at sequentially combine a recurrent neural network with a quadratic program to determine the optimal current flows that must leave the sources towards the distributors before reaching buildings according to their hourly electric needs while minimizing power losses. To achieve this, we proposed a protocol in which we trained a recurrent neural network of type long short term memory with historical electric power needs of buildings. Then, we used the predictive model obtained to forecast the needs of each building during each hour of a day according to historical data. We provide the predicted amount of energy for each building by minimizing the losses in the electrical network by solving the elaborated quadratic program.

The rest of this paper is organized as follows: in Sect. 2, we present some related works. A mathematical modeling of the optimization used to build our protocol, is presented in Sect. 3. Our sequential hybridization is presented in Sect. 4. Results of simulations are presented in Sect. 5. A conclusion ends the paper.

2 Literature Review

Over the years, many works based on optimization techniques for the reduction of power losses in an electrical power distribution system have been proposed. One of the first works that tackled the problem of optimizing the electricity supply was carried out A. Merlin and H. Back [3]. They developed an algorithm that uses the sequential branch opening method. To find an optimal solution, they apply an exhaustive search method of the separation and evaluation type. This method makes it possible to obtain globally optimal solutions. But, it is unanimously recognized that this method requires a prohibitive computing time. Atwa and et al. [4], proposed a methodology to optimally allocate different types of renewable distributed generation units in a distribution system in order to minimize annual energy losses. The planning problem was formulated as nonlinear mixed integer programming (MINLP), with an economic function to minimize the annual energy losses of the system. Being based on an exact method, this method is time consuming for computing the solution of the optimization problem. For her part, Laouafi and al. [5] proposed a solution to solve the problem of the optimal distribution of reactive power in order to reduce power losses in the lines by applying several metaheuristics such as: improved genetic algorithms, particle swarm optimization, simulated annealing, differential evolution and two hybrid formulations. The first is a hybrid formulation between particle swarm optimization and enhanced genetic algorithms with real coding. The second is a hybrid formulation between the differential evolution algorithm (based on a population of solutions) and simulated annealing (based on a single solution). Also, Abdelmoumene and al. [2] proposed an efficient differential evolution (DE) algorithm for the solution of the optimal reactive power dispatch (ORPD) problem. The main objective of their study to minimize the total active power loss with optimal setting of control variables. Moreover, Lenin and al. [6] proposed a prompt particle swarm optimization (PPSO) algorithm to solve the optimal reactive power dispatch (ORPD) Problem. The PPSO algorithm is obtained by combining PSO and the Cauchy mutation and an evolutionary selection strategy. In their approach, they formulated the problem as a nonlinear constrained single-objective optimization problem where the real power loss and the bus voltage deviations are to be minimized separately. Belhour and Abdelouhab [7] have proposed a system for optimizing the distribution of hybrid renewable energies. Their work aimed to determine the best combination of several energy resources to satisfy an energy demand for a specific region. To solve this problem, they opted to model this resource distribution problem in terms of a linear program whose associated resolution method is linear programming. However, this solution is limited as long as it does not take into account variations in demand as a function of time. Kumar and Goyal [8] proposed an implementation of new algorithm Particle Swarm Optimization (PSO) for Energy Saving through minimizing power losses. The objective was to optimize the reactive power dispatch with optimal setting of control variables without violating inequality constraints and satisfying equality constraint. More recently, Jebaraj and Sakthivel [9] proposed a new method of optimization based on swarm intelligence named Sparrow Search Algorithm (SSA) to resolve the Optimization of Power Flow (OPF) problem. The percentage reduction of fuel rate, active transmission loss and deviation in voltage, were examined to show that the approach produces good results.

All the above solutions have the limitation that they do not integrate the variations of consumer needs. However, these variations are real and must be taken into account. All of the results thus presented show that the researchers did not include variations in energy demand.

Chemingui and et al. [10] proposed a Deep Reinforcement Learning agent for controlling and optimizing a school building's energy consumption. It is designed to search for optimal policies to minimize energy consumption, maintain thermal comfort, and reduce indoor contaminant levels in a challenging 21-zone environment. After simulation, the proposed methodology achieved a 21% reduction in energy consumption, a 44% better thermal comfort, and healthier CO_2 concentrations over a one-year simulation. The main limitation of this approach is that the result obtained are theoretical since the agent wasn't deployed into a real school environment to investigate its performance. Also, Jain and et al. [11] proposed a neural network based optimization for building energy management and climate control. Using historical data from the building automation system and the weather station, the authors learn different neural networks that predict energy usage and zone temperatures, and then set up optimization for energy management that allows us to trade-off between energy usage and temperature setpoint tracking. The main limit of this approach is that the proposed protocol doesn't integrate the continual learning of neural networks as long as building properties and weather conditions change with time. In order to engage fully in the field of automatic optimization of the electricity supply, we propose in this paper to combine a method of machine learning with one of the methods used in the literature in order to obtain satisfactory results.

3   Modeling of the Optimization Problem

3.1   Assumptions

The solution developed is based on a number of assumptions:

- The electricity distribution network is built and operational.
- There are several different energy sources and the voltage provided by all energy sources is the same and is 220V.
- Each source produces energy independently and continuously;
- All buildings do not necessarily have the same power need;
- A building's needs can change over time.
- To reach consumers, electricity must pass through one or more splitters.
- Each splitter has the ability to automatically set the amount of current flowing on a cable at any given time.
- The cables leaving the sources to the splitter are identical and therefore have the same resistance. The same is true for the cables leaving the splitter to the consumer buildings.

3.2   Modeling of the Electrical Network

Let us consider a school campus made up of several buildings each having different electrical power needs. Those power needs have been hourly recorded in a file. This campus is powered by energy from different sources of electrical energy production. Before reaching each building, the energy produced by these different sources must pass through a central node which will be called **distributor** which has the capacity to supply electric current at a defined voltage. Its role is to receive the energy from the production sources and provide each of the buildings according to their needs at the moment. We model the campus electrical network with a graph on which each edge represents cables use for electricity transport (see Fig. 1). On that figure, each cable reacts like a resistance when electric current crosses it. Also each node on the left corresponds to the sources of electricity production, the node in the middle to the distributor and those on the right are the buildings of the campus.

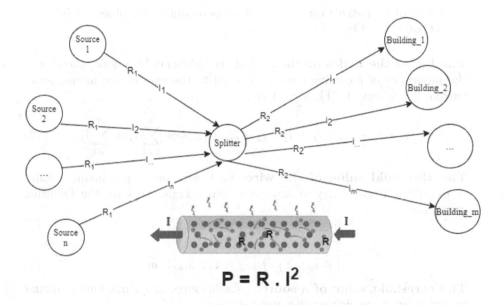

$$P = R \cdot I^2$$

Fig. 1. Campus electrical network.

3.3 Quadratic Optimization Modeling

Considering the situation made in the previous section and that all the cables connecting the various production nodes to the splitter have the same current resistance (R1) and the cables connecting the central node to the consumers have the same resistance (R2).

Objective function: the objective function which is the expression of the power dissipated by Joule effect is written as the sum of the losses on each power line of the network :

$$P = R_1 . I_1^2 + R_1 . I_2^2 + \ldots + R_1 . I_n^2 + R_2 . i_1^2 + R_2 . i_2^2 + \ldots + R_2 . i_m^2$$

$$\Leftrightarrow P = R_1 . \sum_{i=1}^{n} I_i^2 + R_2 . \sum_{j=1}^{m} i_j^2$$

where n is the number of sources and m the number of buildings.

Thus we have the following objective function:

$$P = R_1 . \sum_{i=1}^{n} I_i^2 + R_2 . \sum_{j=1}^{m} i_j^2 \tag{1}$$

Expression of Constraints: The constraints to which this objective function is subject are defined by:

- **The law of the nodes** (at the level of the splitter): At a node, the sum of the intensities of incoming current is equal to the sum of the intensities of current which leave it. Thus, we have:

$$I_1 + I_2 + \ldots + I_n = i_1 + i_2 + \ldots + i_m \Longleftrightarrow \sum_{i=1}^{n} I_i = \sum_{j=1}^{m} i_j \tag{2}$$

- **The threshold value of the wire**: Each wire has a maximum amount of electricity it can carry at any given time. Thus, we have the following constraint:

$$0 \le I_i \le S_i, \; for \; i = 1, 2, 3, \ldots, n \tag{3}$$

$$0 \le i_j \le S_j, \; for \; j = 1, 2, 3, \ldots, m \tag{4}$$

- **The threshold value of a source** : Each source has a maximum amount of electricity it can deliver at any given time.

$$I_i \le T_i, \; for \; i = 1, 2, 3, \ldots, n \tag{5}$$

where I_i represents the current leaving source i at any given moment, T_i represents the maximum value of electric current that can leave source i at any given moment. S_i and S_j represent the maximum current intensity that can flow through the cables without destroying them.

By putting the Eqs. (1), (2), (3), (4) and (5) together, we obtain the following quadratic optimization model :

$$Minimize \; P = R_1 . \sum_{i=1}^{n} I_i^2 + R_2 . \sum_{i=1}^{m} i_i^2$$

Subject to :

$$\sum_{i=1}^{n} I_i = \sum_{j=1}^{m} i_j$$

$$0 \leq I_i \leq S_i, \ pour \ i = 1, 2, 3, \ldots, n$$

$$0 \leq i_j \leq S_i, \ pour \ j = 1, 2, 3, \ldots, m$$

$$I_i \leq T_i, \ for \ i = 1, 2, 3, \ldots, n$$

4   Description of Our Protocol

Our contribution is based on a sequential hybridization of a network of recurrent neurons made up of Long Short Term Memory (LSTM) cells by quadratic programming to achieve the goal of providing electricity needs and reducing power losses. This hybridization is presented by the Fig. 2. In this hybridization, the prediction results of the recurrent neural network are used as initial values of the decision variables of the quadratic program in order to achieve the general objective which is to determine the current flows that must circulate from the sources to the distributor and which minimize the joule losses while respecting the stated constraints.

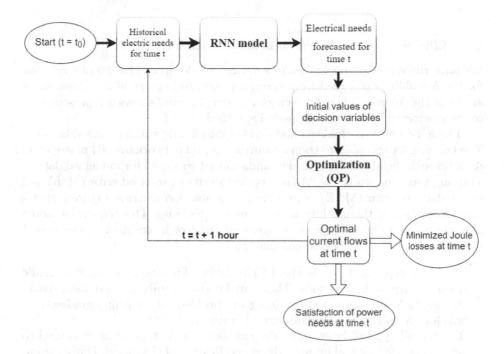

Fig. 2. Modeling of our sequential hybridization.

- After collecting historical electricity consumption data from the different buildings, we use them to train a prediction model that will be used later to predict the needs of each building on an hourly basis.
- At a time t, we retrieve the story needs for this time, which are inputed to the regression model to forecast buildings' needs for that time.
- The values predicted by the regression model for time t are then used to initialize the decision variables of the quadratic program.
- Starting from the initial value of the objective function, we seek the values of the decision variables which make the value of the objective function as small as possible while respecting the constraints which are subjected to them by solving the quadratic program. This makes it possible to obtain the optimum current flows in the electrical network, so as to satisfy user requirements and minimize the joule losses of the electrical network.

5   Simulations and Results

The simulation environment for conducting our experimentation is Jupyter Notebook on an DELL computer with an Intel Core i5 CPU *8thGen* @ 1.90GHz × 8 and 16 GiB of RAM, running Windows 10 Pro. The first part of the experimentation aims to show the efficiency of our protocol. So we have used several parameters to evaluate it. The number of sources is 05, the number of consumer buildings is set to 05. The number of splitter in our electric distribution network is 01.

5.1   Efficiency of the Machine Learning Model

We have run our experiments using a dataset made up of 116,189 observations for the 5 buildings collected from december 2005 to january 2018. Those datas are from the American hourly electricity power. Figure 3 shows a representation of the electricity usage over the period specified.

These data were normalized and divided into 2 sets: training and validation. The training set was used to train a recurrent neural network over 64 passes of its data through the algorithm and the validation set was used for model validation. Training was done using the "Adam" optimizer, mean squared error (MSE) and mean absolute error (MAE) as performance evaluation metrics. Figure 4, shows the information of the machine learning model made up. This sequential model is made up of 3 different types of layers for the five layers making our neural network with a total of 86 910 parameters.

- The first type of layer is the LSTM layer. This type of layer is made up of recurrent LSTM cells. They can be successfully trained using back-propagation over time, thus avoiding the problem of vanishing gradients. It was for layers 1, 2 and 3 of our neural network.
- The second type of layer is the dropout layer. This type of layer is used to prevent overfitting and improve the generalization of the model. During training, randomly selected neurons (along with their connections) are "dropped

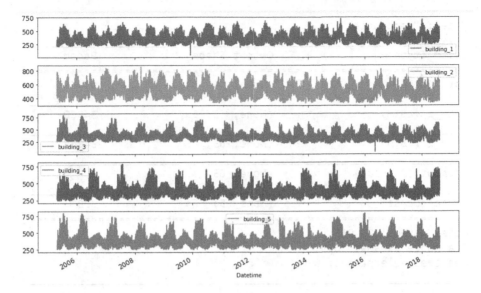

Fig. 3. Representation of observations over time.

out" or temporarily ignored, meaning they are excluded from the forward pass and backward pass computations. This prevents neurons from becoming overly specialized and reduces their interdependence, forcing the network to learn more robust and generalizable features. This type of layer was used for layer 4 of our neural network.

– The third type of layer is the dense layer. Also known as a fully connected layer, the dense layer is a fundamental building block that connects each neuron in one layer to every neuron in the subsequent layer. It's called "dense" because every neuron in a dense layer receives input from all neurons present in the previous layer. This type of layer was used for layer 5 of our neural network.

Figure 5, shows the evolution the mean square error of the model during the training and the validation phases. On that figure we clearly see that at the beginning of the training, the MSE was worth 0.03 and while tending towards the 64th and last, it stabilized at 9.3191×10^{-04} for the training and at 7.1874×10^{-04} for the validation.

Figure 6, shows the evolution the mean square error of the model during the training and the validation phases. On that figure we clearly see that at the beginning of the training, the MSE was worth 0.15 and while tending towards the 64th and last, it stabilized at 0.0220 for the training and at 0.0187 for the validation. Which is good enough because for a regression model, the closer the mae is to zero, the better the model will perform.

After analyzing these different metrics, we are convince that our regression model is able to perform acceptable predictions and can be used to simulate an electrical network.

```
Model: "sequential"
```

Layer (type)	Output Shape	Param #
lstm (LSTM)	(None, 13, 65)	18460
lstm_1 (LSTM)	(None, 13, 65)	34060
lstm_2 (LSTM)	(None, 65)	34060
dropout (Dropout)	(None, 65)	0
dense (Dense)	(None, 5)	330

```
Total params: 86,910
Trainable params: 86,910
Non-trainable params: 0
```

Fig. 4. Layers of the machine learning model.

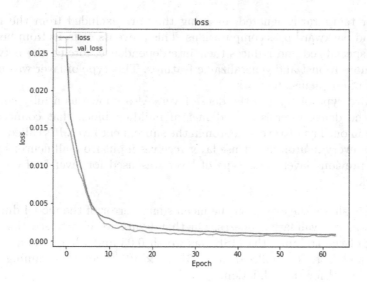

Fig. 5. Mean Square Error per epoch.

5.2   Prediction of Building Needs

By considering configurations stated at the beginning in Sect. 5.1. We forecasted needs of the buildings and the results of this forecasting are presented by the Fig. 7 on which, the blue points represent the historical needs of each building

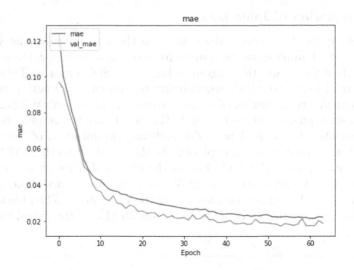

Fig. 6. Mean Absolute Error per epoch.

and the orange dots the needs predicted by our regression model. By using the predicted values we initialized the decision variables of our quadratic program, that is the current flows which must leave the sources towards the distributor.

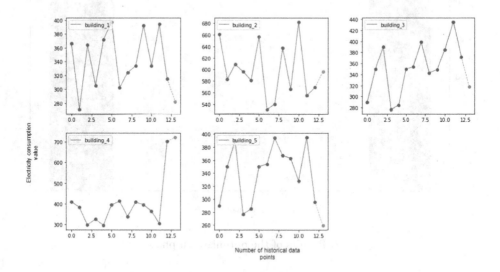

Fig. 7. Forecasting's results.

5.3   Optimization of Joule Losses

After obtaining the forecasted values, we used them to initialize the values of
the currents which must leave the sources towards the distributor. Once this was
done, we solved the quadratic program using the SLSQP method. This method
uses Sequential Least Squares Programming to minimize a function of several
variables with any combination of bounds, equality and inequality. The result of
the optimization phase is given by Fig. 8. On that figure the red bar represents
power losses after the predictions of demands and initialization of the decisions
variables (before the optimization phase). At that level, it worth 80,16 W. After
the optimization phase, the Joule loss in the electrical network is represented
by the green bar and amounts to 56.02 W. Thus, the program will have made it
possible to reduce the Joule power loss by 24.139 W in order to be able to satisfy
the building needs at this time. This represents 30.11% of the initial loss.

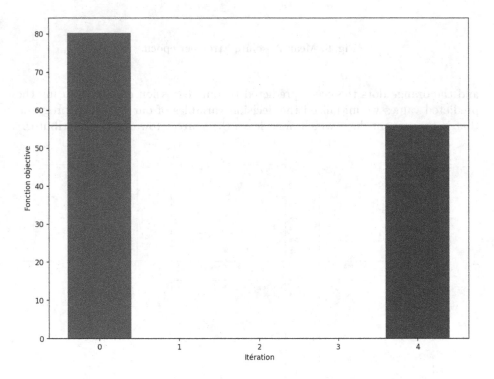

Fig. 8. Results of the optimization phase.

5.4   Comparison with Existing Solution

To show the high efficiency of our protocol, we compared its performance to one
of the best state of art protocol. To make a good comparison, we have used the

same parameters which is the percentage of loss reduction in the network. The result of the comparison is presented in the Table 1. The comparison focused on the percentage of loss reduction by the different experiments since authors did not use the same dataset.

Table 1. Comparison of approaches

Elements	Our approach	Laouafi's approach [5]	Atwa's approach [4]
Variation in needs	Integrated	-	-
Loss Reduction Rate	30,11% for one hour	15.90% compared to the initial case	5% at most per year

6   Conclusion and Future Works

In this paper, we study the problem of reducing poser losses in an electric network. We proposed a hybrid approach according to which we predict the needs of consumers and we use the predicted needs to initialize the decision variables of an optimization model that we solve to determine the current intensities that will make it possible to satisfy the needs. Predicted while reducing power losses due to the Joule effect. This approach makes it possible to better manage the case of reduction of power losses in a network as long as for each moment of a day, it makes it possible to obtain the optimal configuration of the flows of current which make it possible to satisfy each consumer and reduce the energy dissipated by Joule effect.

The following are exciting directions for future research. First, we plan to integrate realities faced during the transport of electricity since the network studied is that of an electricity distribution. We will improve our protocol in order to take into those realities during the phase of optimization. Second, investigate the continual learning of neural networks using model-based reinforcement learning (RL). As building properties and weather conditions change with time, the goal is to minimize the maintenance of neural networks required in manual functional tests by leveraging the exploration capabilities in RL.

Acknowledgment. The authors received no specific funding for this work.

Disclosure of Interests. The authors declare that they have no conflicts of interest related to this research.

References

1. Ahmed, M., Osama, M.: Real-time energy management scheme for hybrid renewable energy systems in smart grid applications. Electric Power Syst. Res. **96**(6), 133–143 (2013)
2. Messaoudi, A., Mohamed, B., Azoui, B.: Optimal reactive power dispatch using differential evolution algorithm with voltage profile control. Int. J. Intell. Syst. Appl. **5**(10), 28 (2013)

3. Merlin A, Back H : Search for a minimal-loss operating spanning tree configuration in an urban power distribution system. In: Proceedings of 5th Power System Computation Conference, pp. 1–18. Cambridge, UK (1975)
4. Atwa, Y.M., El-Saadany, E.F., Salama, M.M.A., Seethapathy, R.: Optimal renewable resources mix for distribution system energy loss minimization. IEEE Trans. Power Syst. **25**(1), 360–370 (2009)
5. Farida, L., Ahcene, B., Salah, L.: A hybrid formulation between differential evolution and simulated annealing algorithms for optimal reactive power dispatch. TELKOMNIKA (Telecommun. Comput. Electron. Control) **16**(2), 513–524 (2018)
6. Lenin, K., Ravindranath, R.B., Surya, K.M.: Prompt particle swarm optimization algorithm for solving optimal reactive power dispatch problem. Int. J. Comput. Technol. Electron. Commun. **2**(4), 1–7 (2013)
7. Souad, B., Abdelouhab, Z.: Optimal repartition of hybrid renewable energy Pv and wind systems. In: 2014 Global Summit on Computer & Information Technology (GSCIT). IEEE, pp. 1–5 (2014)
8. Suresh, K., Kumar, G.S.: A particle swarm optimization for reactive power optimization. Int. J. Comput. Eng. Res. (IJCER) **4**(11), 40–47 (2014)
9. Luke, J., Sithankathan, S.: A new swarm intelligence optimization approach to solve power flow optimization problem incorporating conflicting and fuel cost based objective functions. e-Prime-Adv. Electr. Eng. Electron. Energy **2**(11), 100031 (2022)
10. Yassine, C., Adel, G., Omar, E.: Reinforcement learning-based school energy management system. Energies **13**(23), 6354 (2020)
11. Achin, J., Francesco, S., Enrico, R., Alessandro, D.I., Morari, M.: NeurOpt: neural network based optimization for building energy management and climate control. Learning for Dynamics and Control. PMLR **20**(6), 445–454 (2020)

A New Hybrid Algorithm Based on Ant Colony Optimization and Recurrent Neural Networks with Attention Mechanism for Solving the Traveling Salesman Problem

Anderson Nguetoum Likeufack[ID] and Mathurin Soh[(✉)][ID]

University of Dschang, Dschang, Cameroon
mathurinsoh@gmail.com

Abstract. In this paper, we propose a hybrid approach for solving the symmetric traveling salesman problem. The proposed approach combines the ant colony algorithm (ACO) with neural networks based on the attention mechanism. The idea is to use the predictive capacity of neural networks to guide the behaviour of ants in choosing the next cities to visit and to use the prediction results of the latter to update the pheromone matrix, thereby improving the quality of the solutions obtained. In concrete terms, attention is focused on the most promising cities by taking into account both distance and pheromone information thanks to the attention mechanism, which makes it possible to assign weights to each city according to its degree of relevance. These weights are then used to predict the next towns to visit for each city. Experimental results on instances TSP from the TSPLIB library demonstrate that this hybrid approach is better compared to the classic ACO.

Keywords: Traveling Salesman Problem · Neural Networks Hybridization · Attention Mechanism · Ant Colony Algorithm

1 Introduction

In the field of robotics, motion and path planning play a crucial role in accomplishing complex tasks. Robots often need to perform efficient and optimized movements to reach different points of interest, collect information, or interact with their environment. Solving the Traveling Salesman Problem (TSP) in robotics can improve the operational efficiency of robots by reducing travel times and unnecessary movements. However, the TSP is considered an NP-hard problem [1,4], meaning that there is no algorithm that can solve it in a reasonable amount of time.

Faced with this complexity, researchers have proposed several methods to obtain an approximate solution to the TSP. One of these methods is the Ant Colony Optimization (ACO) metaheuristic proposed by Marco Dorigo [3],

© The Author(s), under exclusive license to Springer Nature Switzerland AG 2024
P. Melatagia Yonta et al. (Eds.): CRI 2023, CCIS 2085, pp. 143–153, 2024.
https://doi.org/10.1007/978-3-031-63110-8_12

inspired by experiments on the behavior of real ants [2]. Although effective in solving small instances, it has limitations in exploring the solution space, making it inefficient for large-scale TSP instances. One approach to overcome this limitation would be to combine ACO with neural networks with attention mechanisms to intelligently guide the ants in promising areas. In this context, we propose in this paper a new hybrid approach called ACO-RNN (Ant Colony Optimization-Recurrent Neural Network) to enhance ACO exploration by introducing a novel update formula for the pheromone matrix.

In the rest of this document, we provide a formal description of the TSP and its applications in the second section. We briefly review the state of the art in TSP resolution in the third section. Next, we explore the hybridization techniques used in our approach in the fourth section. In the fifth section, we provide a detailed description of the ACO-RNN approach we developed. Finally, in the last section, we present the results of our experiments.

2 Formal Description of the TSP

The Traveling Salesman Problem involves finding an optimal tour for a traveling salesman who wants to visit exactly n cities and return to the starting city [1,4]. Mathematically, the TSP can be formulated as follows:

Let $G = (V, E)$ be a complete undirected graph, where V represents the set of cities to be visited, and E represents the set of edges connecting the cities. Each edge $(i, j) \in E$ is associated with a distance or cost $d(i, j)$ representing the distance between cities i and j. The objective of the TSP is to find a Hamiltonian cycle of minimum length in the graph G, i.e., a path that visits each city exactly once and returns to the starting city. The total length of the cycle is given by the objective function.

$$\text{Minimize} \quad \sum_{(i,j)\in E} d(i,j)x(i,j)$$

$$\text{Subject to} \quad \sum_{j\in V} x(i,j) = 1, \quad \forall i \in V$$

$$\sum_{i\in V} x(i,j) = 1, \quad \forall j \in V$$

$$\sum_{i\in S}\sum_{j\in V\setminus S} x(i,j) \leq |S| - 1, \quad \forall S \subset V, |S| \geq 2$$

$$\text{x(i, j)} \in 0, 1, \quad \forall(i,j) \in E$$

The TSP has attracted significant interest from the scientific community due to its applications beyond robotics. It finds applications in various other domains, such as logistics for optimizing delivery routes, bioinformatics for analyzing DNA sequences and finding common motifs, electronic circuit design for optimizing wire paths in integrated circuits, etc.

3   State of the Art

The TSP has been the subject of extensive research for many years, leading to the proposal of multiple methods and algorithms to solve it. In this section, we will focus specifically on hybrid techniques and those using deep learning to solve the TSP.

Hybridization involves combining features from two different methods to leverage the advantages of both. Hybrid metaheuristic algorithms were introduced in the works of Glover [5], J. J. Grefenstette [6], and Mühlenbein et al. [7]. According to the classification of hybridization methods proposed in [8], the hierarchical classification is characterized by the level and mode of hybridization [9]. The level of hybridization can be either low-level or high-level. In low-level hybridization, a metaheuristic replaces a specific operator of another method that encompasses it. On the other hand, in high-level hybridization, each metaheuristic retains its own characteristics throughout the hybridization. Each level of hybridization gives rise to two cooperation modes: the relay mode and the co-evolutionary mode. In the relay mode, methods are executed sequentially, where the result of the first method becomes the input to the next method. When different methods work in parallel to explore the search space, it is called the co-evolutionary mode [9]. In this context, Ant-Q was developed by Gunter Stützle and Holger H. Hoos in 1996 [10], as an improvement of the Ant System. Ant-Q combines ACO with q-learning technique by integrating a learning mechanism based on a Q-value table. This allows it to combine pheromone-based exploration with reward-based exploitation, thus improving the search performance and adaptability of the algorithm. Pu Y-F et al. [11] propose a new conceptual formulation of the Fractional Order Ant Colony Algorithm (FACA), which is based on fractional long-term memory. This approach aims to improve the optimization capability of traditional ant colony algorithms by modifying the transition behavior of the ants. Similarly, a cooperative parallel hybrid approach proposed in the article [12] by Gulcu was developed by combining the ACO algorithm to generate solutions in parallel with the 3-Opt local search heuristic to improve these solutions. Similarly, PSO-ACO-3Opt [13] uses the PSO algorithm to adjust ACO parameters and integrates the 3-Opt technique to avoid premature convergence of ACO. Xiaoling et al. propose a new hybrid approach based on the Adaptive State Slime Mold (SM) organism model and the Fractional Order Ant Colony System (SSMFAS) [14] to tackle the TSP.

The approach developed in this paper uses a similar process to Ant-Q [10], with the difference being that it is based on pheromone exploration combined with predictions from a neural network using the attention mechanism. With the advent of artificial intelligence, some researchers have been using deep learning techniques to solve the TSP. Notably, the work of Bresson et al. proposes a new approach called TSP Transformer [15]. This model is based on a classic Transformer encoder with multi-head attention and residual connections, but it uses batch normalization instead of layer normalization. In this approach, decoding is performed in an auto-regressive manner, and a self-attention block is introduced in the decoder part. Minseop et al. [16] propose a new

CNN-Transformer model based on partial self-attention, where attention is computed only on the recently visited nodes in the decoder. This is because the linear embedding in the standard Transformer model does not take into account local spatial information and has limitations in learning local compositionality.

In this paper, a similar idea is employed by combining the Ant Colony Optimization (ACO) process with the results of predictions from a neural network using the attention mechanism. This hybrid approach, named ACO-RNN, aims to leverage the predictive power of neural networks to intelligently guide the ants in exploring promising areas while optimizing the pheromone-based exploration process. By combining the strengths of both ACO and neural networks with attention mechanisms, the proposed approach seeks to improve the exploration efficiency and overall performance for solving the TSP.

4   Principle of Used Hybridization Techniques

4.1   Principle of ACO

The classical Ant Colony Optimization (ACO) proceeds as follows [3]:

1. Initialization: Place each ant on a city randomly.
2. Construction of solutions: Each ant moves from one city to another following specific rules. The probability of choosing a particular city depends on the amount of pheromone deposited on that city and the distance between cities.
3. Pheromone update: After all ants have constructed their solutions, the amount of pheromone deposited on each edge is updated. Ants deposit more pheromone on the edges of higher-quality solutions. The update of the pheromone matrix is done using the formula:

$$\tau_{ij} = (1 - \rho) \cdot \tau_{ij} + \sum_{k=1}^{m} \Delta \tau_{ij}^{k} \tag{1}$$

where the symbols have the following meanings:
- τ_{ij} represents the amount of pheromone on the edge connecting vertex i to vertex j,
- ρ is the pheromone evaporation rate,
- m is the number of ants in the colony,
- $\Delta \tau_{ij}^{k}$ is the amount of pheromone deposited by ant k on the edge ij,

4.2   Autoencoder Model with Attention Mechanism

In this architecture, the encoder is constructed using an LSTM layer that returns sequences. Then, the attention mechanism is applied to the encoder's output to highlight important parts of the sequence. Next, the decoder uses another LSTM layer to generate the reconstructed output. The overall model is created by connecting the input to the outputs of the encoder and decoder (Fig. 1).

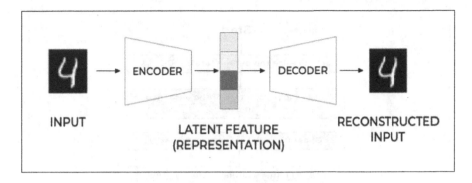

Fig. 1. General structure of an autoencoder [18]

5   Hybridization of Ant Colony Optimization and Neural Networks with Attention Mechanism

5.1   Basic Idea

The fundamental idea of our approach, which is a low-level co-evolutionary hybridization, is to guide the behavior of ants using predictions from a neural network with an attention mechanism. Traditionally, ACO uses pheromones to guide ants towards neighboring cities and explore the solution space. However, this approach may be limited in terms of efficiency and quality of solutions obtained. The objective of this approach is to use a neural network to predict the next cities to visit based on the best solution obtained by ACO so far. This approach is divided into two main steps.

5.2   Resolution Steps

1. **Model Training**

   As part of training the model, we use ACO to generate initial solutions. We randomly place ants on cities that will perform cuter a full turn. Once these solutions appear, we encode the best solution into a data format suitable for training the model using One-Hot(per position) encoding. One-hot encoding [17], This allows us to represent the solution in a structured and coherent way, thus making it easier to learn the model (Fig. 2).

   To illustrate how we use one-hot encoding, assume a 5-city TSP problem with solution [2, 0, 3, 1, 4]. To encode this solution using one-hot encoding, each city will be represented by a binary vector of size 5, where all values are zero, except the one corresponding to the city index, which will be one. Here is the one-hot encoding for the given TSP solution:

   - **City 2:** [0, 0, 1, 0, 0]
   - **City 0:** [1, 0, 0, 0, 0]
   - **City 3:** [0, 0, 0, 1, 0]
   - **City 1:** [0, 1, 0, 0, 0]

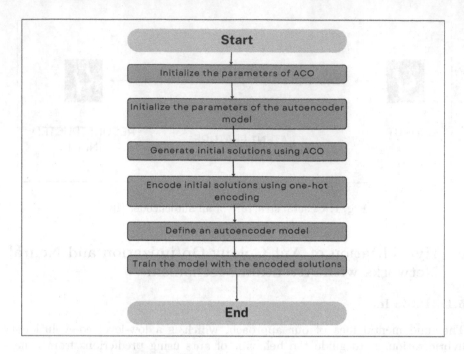

Fig. 2. Algorithm flowchart for model training phase

– **City 4:** $[0, 0, 0, 0, 1]$

After representing each city in the TSP solution as a distinct binary vector. We form a tensor by concatenating these one-hot vectors into an ordered sequence, where each vector represents the state of a city at a certain time. In our example, this tensor will be:

$[\ [0, 0, 1, 0, 0], [1, 0, 0, 0, 0], [0, 0, 0, 1, 0], [0, 1, 0, 0, 0], [0, 0, 0, 0, 1] \ ]$.

Then we train a recurrent neural network with attention mechanism on this sequence of one-hot vectors to allow the model to learn the sequential relationships between cities and capture the dependencies and relationships between cities. Instead of treating all cities in the sequence uniformly, the attention mechanism allows the model to give more weight or importance to cities that are deemed more relevant for predicting the next city to visit.

2. **Finding the Solution**

Finding the final solution involves the iterative execution of ACO. At each iteration, the update of the pheromone matrix is carried out based on the predictions provided by the model. This approach combines the use of ACO to explore the search space and the model's ability to provide predictions based on the information learned during training. As a result, the update of the pheromone matrix takes into account both the knowledge of ACO and the information provided by the model, allowing for a more effective exploitation of both approaches for finding the optimal solution. The new formula for updating the pheromone matrix that we propose is as follows:

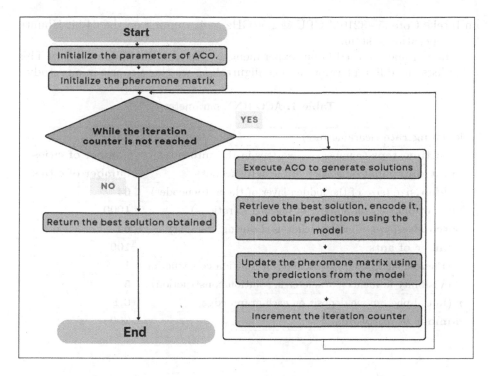

Fig. 3. Algorithm flowchart for ACO-RNN solution search phase

$$\tau_{ij} = (1 - \rho) \cdot \tau_{ij} + \text{predictions}[i, j] \qquad (2)$$

where the symbols have the following meanings:
- τ_{ij} represents the amount of pheromone on the edge connecting vertex i to vertex j,
- ρ is the pheromone evaporation rate,
- $predictions[i, j]$ corresponds to the probability predicted by the model between city i and j .

This integration of the model's results into the pheromone update process allows us to leverage the model's prediction capabilities to guide the search towards more promising solutions and potentially improve the performance of ACO in solving the TSP. The Fig. 3 illustrates the process of the final search for the solution in our approach.

6   Experiments

6.1   Implementation Environment

The experiments were conducted using the Python programming language on a specific machine. The machine used had a configuration with 6 GB of RAM and an Intel®Core i5−3210M CPU @ 2.50 GHz × 4, running on the 64-bit Ubuntu 20.04 operating system.

The parameters used in our experiment are listed in the Table 1 below. This table lists the different values and configurations we chose to use in our study.

Table 1. ACO-RNN parameters

learning_rate (learning rate)	0.01
num_timesteps (number of timesteps in the autoencoder)	number of cities
input_dim (input dimension of the autoencoder)	number of cities
hidden_size (size of the hidden layer of the autoencoder)	64
num_epochs (total number of training iterations)	1000
batch_size (size of mini-batches used during training)	64
number of ants	100
α (Pheromone influence parameter in solution construction)	1
β (Visibility influence parameter in solution construction)	5
τ (Initial pheromone deposit on each graph edge)	0.1
number of iterations	200

6.2   Results, Analysis and Discussion

In this section, we present and interpret the results obtained in the context of our study. We compare the performance of ACO-RNN with the classical ACO.

We will begin by evaluating the different errors between our approach, the ACO, and the known solutions on small Fig. 4 and medium-sized Fig. 5 problem instances. In order to compare the performance of these different approaches, we will use the following formula to calculate the error rates:

$$\text{Error Rate} = \left(\frac{\text{Distance of the known solution} - \text{Distance of the approach}}{\text{Distance of the known solution}} \right) \times 100 \quad (3)$$

In this formula, the distance of the approach refers to the total length of the path found by our method or the ACO, while the distance of the known solution represents the total length of the path for the best-known solution.

According to the results presented in Fig. 4 and Fig. 5 in red color, ACO-RNN demonstrates significant superiority over the classical ACO, producing results with considerably reduced errors. This improvement is attributed to more effective exploration by ACO-RNN compared to classical ACO, leveraging the quality of predictions provided by the neural network model.

We have grouped the comparisons between our approach and the known solutions for small and medium-sized instances respectively in Table 2 and Table 3 below. These tables present the distances calculated by our approach and the distances of the known solutions (Table 2 and Table 3).

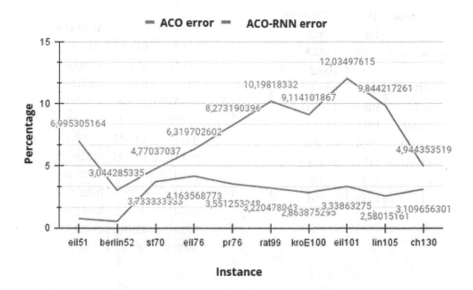

Fig. 4. Comparative error curves of ACO and ACO-RNN in terms of cost on small instances (Color figure online)

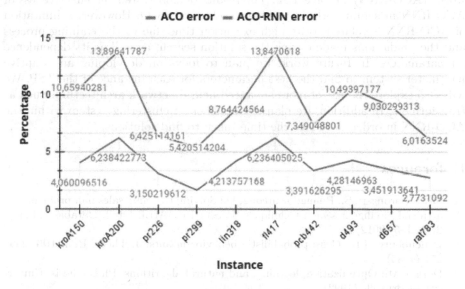

Fig. 5. Comparative error curves of ACO and ACO-RNN in terms of cost on medium instances (Color figure online)

Table 2. Small Instance Comparisons

Instance	BKS	ACO-RNN
eil51	426	429.3
berlin52	7542	7583.8
st70	675	700.2
eil76	538	560.4
pr76	108159	112000.1
rat99	1211	1334.5
rd100	7910	8036.0
kroA100	21747	21554.2
eil101	629	650.8
lin105	14379	14750

Table 3. Medium Instance Comparisons

Instance	BKS	ACO-RNN
ch130	6110	6300.1
pr144	6110	57189.8
ch150	6528	6987.4
kroA200	29368	33449.1
rd400	15281	15900.5
fl417	**11861**	13503.4
pr439	107217	112286.9
pcb442	50778	53850.0
rat575	6773	7053.7
p654	34643	35982.0
u724	41910	44002.5
rat783	8806	9335.8

7 Conclusion

In this study, we proposed a low-level co-evolutionary hybridization approach between the ant colony algorithm and attention mechanism-based neural networks (ACO-RNN) to solve TSP. The results demonstrated the effectiveness of ACO-RNN in significantly improving on classical ACO. However, a limitation of ACO-RNN is related to its high execution time due to the training process and the predictions made during the solution search, as well as its dependence on parameters. In future work, we plan to focus on developing an adaptive parameter system to find the best parameters for each instance of the TSP. We will also explore the use of more advanced neural network architectures such as transformers. In addition, we plan to implement a clustering system within the ACO-RNN in order to reduce the time taken to find solutions.

References

1. Suwannarongsri, S., Puangdownreong, D.: Solving traveling salesman problems via artificial intelligent search techniques. Recent Res. Artif. Intell. Database Manag. **2**(2), 1–5 (2012)
2. Deneubourg, J.L.: Goss: probabilistic behavior in ants. J. Theor. Biol. **105**, 259–271 (1983)
3. Dorigo, M.: Optimization, learning, and natural algorithms. Ph.D. thesis, University of Brussels (1992)
4. Dehan, M.: Distribution of medicinal products derived from blood plasma in Belgium: a case study. Catholic University of Louvain, 50–73 (2018)

5. Glover, F.: Heuristics for integer programming using surrogate constraints. Dec. Sci. 156–166 (1977)
6. Grefenstette, J.J.: Incorporating problem specific knowledge into genetic algorithms. In: Davis, L.D. (ed.) Genetic Algorithms and Simulated Annealing. Pitman, London (1987)
7. Gorges-Schleuter, M., Kramer, O., Mühlenbein, H.: Evolution algorithms in combinatorial optimization. Parall. Comput. 65–88 (1988)
8. Jourdan, L., Dhaenens, C., Talbi, E.G., Gallina, S.: A data mining approach to discover genetic and environmental factors involved in multifactorial diseases. Knowl.-Based Syst. 15(4), 235–242 (2002)
9. Hachimi, H.: Hybridations d'algorithmes métaheuristiques en optimisation globale et leurs applications, Ph.D. thesis, Université Mohammed V - Agdal, Rabat, Institut National des Sciences Appliquées de Rouen, Rabat, Maroc; Rouen, France, 2013. Spécialité: Mathématiques appliquées et Informatique. Option: Optimisation, Analyse numérique, Statistique
10. Gambardella, L.M., Dorigo, M.: Ant-Q: A Reinforcement Learning Approach to the Traveling Salesman Problem, pp. 252–260. Morgan Kaufmann, Palo Alto (1995)
11. Pu, Y.F., Siarry, P., Zhu, W.Y., Wang, J., Zhang, N.: Fractional-order ant colony algorithm: a fractional long term memory based cooperative learning approach. Swarm Evol. Comput. 69, 101014 (2022)
12. Gulcu, S., Mahi, M., Baykan, O., Kodaz, H.: A parallel cooperative hybrid method based on ant colony optimization and 3-Opt algorithm for solving traveling salesman problem. Soft. Comput. 22, 1669–1685 (2018)
13. Mahi, M., Baykan, O., Kodaz, H.: A new hybrid method based on particle swarm optimization, ant colony optimization and 3-opt algorithms for traveling salesman problem. Appl. Soft Comput. 30, 484–490 (2015)
14. Gong, X., Rong, Z., Wang, J., et al.: A hybrid algorithm based on state-adaptive slime mold model and fractional-order ant system for the travelling salesman problem. Complex Intell. Syst. (2022)
15. Bresson, X., Laurent, T.: The transformer network for the traveling salesman problem. arXiv preprint arXiv:2103.03012
16. Jung, M., Lee, J., Kim, J.: A lightweight CNN-transformer model for learning traveling salesman problems. arXiv preprint arXiv:2305.01883v1
17. Chieng Hoon Choong, A., Lee, N.K.: Evaluation of convolutionary neural networks modeling of DNA sequences using ordinal versus one-hot encoding method. bioRxiv (2017)
18. Michelucci, U.: An introduction to autoencoders. arXiv preprint arXiv:2201.03898v1 (2022)

Recurrent Neural Network Parallelization for Hate Messages Detection

Thomas Messi Nguele[1,2,3]([✉]), Armel Jacques Nzekon Nzeko'o[1,3],
and Damase Donald Onana[1]

[1] Computer Science Department, University of Yaounde I, FS, Yaounde, Cameroon
{armel.nzekon,donald.onana}@facsciences-uy1.cm,
thomas.messi@facsciences-uy1.cm
[2] Computer Engineering Department, University of Ebolowa, HITLC,
Ebolowa, Cameroon
[3] Sorbonne Université, IRD, UMI 209 UMMISCO, 93143 Bondy, France

Abstract. Hate speech is a threat to democratic values, because it stimulates incitement to discrimination, which international law prohibits. To limit the harmful effects of this scourge, scientists often integrate into social network platforms models provided by deep learning algorithms allowing to detect and react automatically to a message with a hateful nature. One of the particularities of these algorithms is that they are so efficient as the amount of data used is large. However, sequential execution of these algorithms on large amounts of data can take a very long time. In this paper we first compared three variants of Recurrent Neural Network (RNN) to detect hate messages. We have shown that Long Short Time Memory (LSTM) provides better metric performance, but implies more important execution time in comparison with Gated Recurrent Unit (GRU) and standard RNN. To have both good metric performance and reduced execution time, we proceeded to a parallel implementation of the training algorithms. We proposed a parallel implementation based on an implicit aggregation strategy in comparison to the existing approach which is based on a strategy with an aggregation function. The experimental results on an 8-core machine at 2.20 GHz show that better results are obtained with the parallelization strategy that we proposed. For the parallel implementation of an LSTM using the dataset obtained on kaggle, we obtained an f-measure of 0.70 and a speedup of 2.2 with our approach, compared to a f-measure of 0.65 and a speedup of 2.19 with an explicit aggregation strategy between workers.

Keywords: Deep Learning · Recurrent Neural Network · hateful messages recognition system · parallel programming

1 Introduction

Digital communication has been very successful since the advent of the internet, as can be seen in the annual report on digital in 2022[1]. In that report, we

---

[1] https://wearesocial.com/, accessed on 2023/06/09.

learn that, at the beginning of 2022 there were 4.62 billion active users of social networks, i.e. 58.4% of the world's population. These high usage rates are due to the various positive points that this communication channel brings. However, this way of communicating also has a number of drawbacks, such as the dissemination of hateful messages, which are often a source of murder, suicide, incitement to violence or discrimination of any kind. Many works are carried out with the aim of reducing this negative impact of social network platforms.

The enormous amount of data generated by social media platforms is a boon for deep learning algorithms. Indeed, more the quantity of data used is important, more the algorithm is efficient. On the other hand, the sequential execution of these algorithms on these large amounts of data can take a very high execution time [1]. It is possible through an appropriate use of multi-core architectures to reduce this execution time. In this paper, we are interested to identifying hateful messages that can be disseminated on social networks using Recurrent Neural Network (RNN) algorithms. We propose the parallelization of RNN for hateful messages recognition. The objective is to take advantage of multi-core hardware architectures to reduce the execution time of training algorithms while preserving their metrics performance. We firstly compared sequential executions of standard RNN, Long Short Time Memory (LSTM) and Gated Recurrent Unit (GRU) on two datasets. We subsequently implemented each of the training algorithms in parallel, using and comparing two different approaches. One approach performing explicit aggregation (arithmetic average) between the processing units for the update phase and another using mutex software synchronization between processing units. After our experiments on an 08-core machine, we had better performance for the parallelization of an LSTM using the mutex synchronization. We obtain a speedUp of 2.2, a f-measure of 0.70, a recall of 0.69 and a precision of 0.72 with one dataset containing 5190 tweets including 1880 offensive tweets, 1430 hateful tweets and 1880 neither offensive nor hateful.

The rest of this paper is organised as follows: Sect. 2 presents the concepts necessary for the understanding our work. Section 3 presents the existing work related to ours, Sect. 4 present and explain our parallel implementation, Sect. 5 present the current experimental results, and finally Sect. 6 will conclude and give some perspectives.

2   Background

Recreant Neural Network [2] are types of neural networks used for processing sequential data like a text. Unlike the Feed-Forward Neural Network where information only propagates in one direction, from front to back, RNNs are neural networks with recurrent connections and where the information is propagated in both directions. In this article, we use and compare 03 variants of RNN namely: standard RNN, LSTM, and GRU. A standard RNN simulates a discrete-time dynamical system. It is a system with an input x_t, an output y_t and a hidden state h_t formally defined like in Eq. 1, where f_h and f_o are parameterized state transition and output functions respectively. As shown by Pascanu et al. [3] standard RNN have difficulties in practice to be able to handle very long sequences due to the problem of vanishing or exploding gradient. Several solutions have emerged to overcome these anomalies, such as the use of special architectures like LSTM and GRU both based on the use of a gate mechanism.

$$h_t = f_h(x_t, h_{t-1})$$
$$y_t = f_o(h_t) \tag{1}$$

To handling long-term temporal dependencies in the data, an LSTM uses a memory cell (c_t) and 03 gate whose role is to control the flow of information in and out. We have the forgate gate (f_t) which allows to control the quantity of information to keep, the input gate (i_t) allows to select the information to add in c_t and the output gate (o_t) allows to obtain the hidden state h_t as a function of c_t. Formally, to have the hidden state h_t with an LSTM, we proceed as in Eq. 2. Where All W denote the weight matrices and b are the biases (the model parameters). $\bar{c}_t$ designates the intermediate state of the secondary memory c_t (state before being filtered by the gates).

$$f_t = \sigma(W_{xf}x_t + W_{hf}h_{t-1} + b_f)$$
$$i_t = \sigma(W_{xi}x_t + W_{hi}h_{t-1} + b_i)$$
$$\bar{c}_t = tanh(W_{xc}x_t + W_{hc}h_{t-1} + b_c)$$
$$c_t = f_t \odot c_{t-1} + i_t \odot \bar{c}_t \tag{2}$$
$$o_t = \sigma(W_{xo}x_t + W_{ho}h_{t-1} + b_o)$$
$$h_t = o_t \odot tanh(c_t)$$

The sigmoid (σ) and hyperbolic tangent $(tanh)$ activation functions were respectively used to calculate the gates and the hiden state.

The main difference between LSTM and GRU is that, GRU has fewer parameters and use only two gates namely: reset gate (r_t) and update gate (z_t). The reset gate regulates the flow of new input to the previous memory and the update gate determines how much of the previous memory to keep. The calculation of hidden state h_t with a GRU occurs as presented in Eq. 3.

$$z_t = \sigma(W_{xz}x_t + W_{hz}h_{t-1} + b_z)$$
$$r_t = \sigma(W_{xr}x_t + W_{hr}h_{t-1} + b_r)$$
$$\bar{h}_t = tanh(W_{xh}x_t + W_{hh}(r_t \odot h_{t-1}) + b_h) \tag{3}$$
$$h_t = z_t \odot h_{t-1} + (1 - z_t) \odot \bar{h}_t$$

Learning algorithm with RNN like others neural networks is based on iterative execution of two phases, the forward and the back-forward phase. In the forward phase we compute the outputs as show in Eq. 1, 2 and 3, and in the back forward phase we compute the partial derivatives of each parameter of the model to then update them [7]. Algorithm 1 presents the sequential training of

RNN we use for text classification. Gradient descent in its mini-batch version was used as an optimization procedure, choosing the cross-entropy loss as the error function. It is an error function used for multi-class classification problems (as in our case), which allows to quantify the difference between the predicted vector output (y) and the expected one $(\bar{y})$. It is defined as in Eq. 4 for one instance i of the dataset.

$$\ell_i(y, \bar{y}) = -\sum_{j=1}^{n} \bar{y}_j log(y_j) \tag{4}$$

$$\theta = \theta - \frac{\lambda}{M} \sum_{i=1}^{M} \frac{\partial \ell_i}{\partial \theta} \tag{5}$$

Algorithm 1 : Training

Input($(\mathbf{X,Y})$; λ: learning rate; $\mathbf{M}$: batch size; **epoch**: maximal numbers of epoch)
Output Θ : Set of model parameters
Start

1: initializeParameters(Θ)   //initialize the Θ parameters
2: (xtrain, ytrain) (xval, yval) ← splitData(X, Y)
3: B ← getBatch(M,xtrain,ytrain)
4: e ← 1,   stop ← 0,   valLoss ← ∞
5: **while** $e \leq$ epoch **and** stop < 4 **do**
6:    **for** b$_i$ ∈ B **do**
7:       loss ← 0
8:       zeroInitialize(dΘ)   //for each batch, initialize the partial derivatives to zero
9:       **for** (x, y) ∈ b$_i$ **do**
10:          y$_{pred}$ ← Forward(x, Θ) // for each sentence x, predict its class

11:          loss ← loss + lossEntropy(y, y$_{pred}$) //accumulate the prediction error

12:          dΘ ← dΘ + Backforward(Θ, x, y, y$_{pred}$) // accumulate the partial derivatives
13:       **end for**
14:       update(Θ, dΘ, λ, M)   // update the parameters
15:    **end for**
16:    **if** evalModel(Θ, xval, yval) $>$ valLoss **then**
17:       stop ← stop + 1   // if the performance hasn't increased, increment the variable stop
18:    **else**
19:       valLoss ← evalModel(Θ, xval, yval)
20:       stop ← 0   // if not, the stop variable is reset to zero
21:    **end if**
22:    e ← e + 1
23: **end while**
24: **Return** Θ
End

The parameter update is carried out as in Eq. 5, where λ (the learning rate) is a parameter fixed before the start of learning and which may change or not throughout the learning, and M is the batch size. Finally, it is important to specify that the words in natural language constituting the text are first transformed into a vector to be able to be used in the algorithm. For our study, we simply used a word2vec algorithm provided by gensim[2] python library. Considering the

---

[2] https://radimrehurek.com/gensim/models/word2vec.html.

iterative execution of the forward and backforward phase in algorithm 1, and setting: n_e the maximum number of epochs, n_d the number of elements of the dataset, n_h the number of neurons in the hidden layer, n_v the size of the input vector representing a word, n_o the size of the output vector, and n_s the maximum sentence size, then the worst-case time complexity for a standard RNN is $O(n_d n_e(n_o n_h + n_s n_h(n_h + n_v))$ and for LSTM and GRU is $O(n_d n_e(n_s n_h + n_s(n_v + n_h(n_h + n_v)))$. The main goal in this paper is to provide a parallel implementation of this algorithm in order to reduce the execution time and maintaining as much as possible prediction performance.

3 Related Work

Several works have been carried out on the parallel implementation of deep learning algorithms. By analyzing this works, we mainly observe two parallelization approaches, namely model parallelization and data-level parallelization which can be carried out synchronously or asynchronously. For example Martin Abadi et al. [4] who in their work present the Synchronous Parallel Stochastic Gradient Descent (S-PSGD) algorithm. It is a parallel implementation of stochastic gradient descent exploiting data-level parallelization. The algorithm is synchronous in the sense that the processing units synchronize with each other to aggregate their gradient in order to update the global model.

Jeffrey Dean et al. [1] combine model parallelization and data-level parallelization asynchronously. They present the DownPour SGD algorithm. The principle is that each processing unit is responsible for updating a subset of model parameters. This approach is asynchronous in the sense that, the processing units operate independently of each other.

Zhiheng Huang et al. [5] provide a parallel RNN training algorithm for language modeling. For their implementation, each processing unit contains in memory a complete copy of the model, but also a subset of the dataset. The difference with the S-PSGD algorithm is that here, each processing unit updates its local model. The common point of the works that we have just cited is the fact that they all use data-level parallelization. The main difference is made on the strategy used for updating the global model. We propose in this article a parallel implementation strategy with mutex synchronization between processing units to perform updates. We compare this strategy with the one where the processing units update locally and aggregate their results.

4 Recurrent Neural Network Parallelization

4.1 Parallelization Strategy

Stochastic gradient descent (SGD) is the most used optimization procedure when training deep learning model. Unfortunately, its traditional formulation is inherently sequential which makes it difficult to use with very large volumes of data [1]. This is because the time required to traverse the data fully sequentially can be expensive. An effective solution to reduce the training time of a neural network is the use of parallel programming. The main approach to parallelize the training of a neural network is to distribute the computation of the derivatives over several processing units by exploiting data-level parallelization [5]. By following this approach, we propose in this paper a parallel implementation strategy performing a mutex software synchronization between the processing units for update the global model. The principle is that, each processing unit is assigned a local copy and a subset of the dataset (see Fig. 1). They each use their local copy of the model to calculate the derivatives (∇P) of the parameters. After calculating the derivatives, each processing unit one after the other enters in critical section using a mutex variable to update the parameters of the global model accessible from a shared memory.

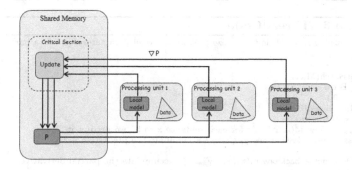

Fig. 1. Parallelization with mutex synchronization. Each data part is assigned to a processing unit which compute the derivatives (∇P), and uses the critical section to update the global model.

This is therefore a parallelization at the data level because each processing unit will execute the same set of instructions simultaneously with the others but with different data. It is synchronous in the sense that, we use software synchronization (mutex) between the processing units for update operations.

4.2 Parallel Algorithm

The implementation was carried out in a shared memory environment and equipped with a multi-core architecture. The processing units then correspond to the different cores of a processor. Each core, through a thread executes simultaneously with the other threads a set of instructions. The main thread executes algorithm 2. It is responsible for initializing the parameters of the global model (line 1). At each epoch (line 4 to 9), it partitions the data into P subsets, where p designates the number of processing units available. It then launches the execution of the slave threads with a copy of the model parameters, as well as a subset of the data sets.

Algorithm 2 : ParallelTraining

Input ((X,Y): set of sentences x and their label y ; λ: learning rate; M: batch size; epoch : numbers of epoch; P: numbers of thread)
Output Θ : set of model parameters
Start

1:   initializeParameters(Θ)
2:   $e \leftarrow 1$,   stop $\leftarrow 0$
3:   **while** $e \leq$ epoch **and** stop < 4 **do**
4:      partition the data (X,Y) into P part d_i
5:      **for** each d_i **do**
6:         ParametersCopy(Θ, Θ_i) // each slave thread will make a local copy of the global model

7:         ThreadCode(d_i, Θ_i, λ, M) // function executed by each thread
8:      **end for**
9:      the model is evaluated as in algorithm 1 (line 17 to 21)
10:     $e \leftarrow e + 1$
11: **end while**
12: Return Θ
End

Algorithm 3 : ThreadCode

Input (λ: learning rate; M: le batch size ; Θ_i: local model parameters, d_i: data partition)
Output
Start

1:   $B \leftarrow$ getBatch(M,d_i)
2:   **for** $b_i \in B$ **do**
3:      loss $\leftarrow 0$
4:      zeroInitialize(dΘ)
5:      **for** $(x, y) \in b_i$ **do**
6:         $y_{pred} \leftarrow$ forward(x, Θ_i)   // for each sentence x in b_i, we predict its class
7:         loss $\leftarrow$ loss + lossEntropy(y, y_{pred}) // accumulate the prediction error

8:         $d\Theta_i \leftarrow d\Theta_i +$ backforward(Θ_i, x, y, y_{pred}) // accumulate the partial derivatives
9:      **end for**
10:     **lockmutex_update**
11:     update(Θ, $d\Theta_i$, λ, M)   // each slave thread update the global model in critical section
12:     ParametersCopy(Θ, Θ_i) //then retrieves the updated global parameters
13:     **freemutex_update**
14: **end for**
End

Each slave thread executes algorithm 3 simultaneously with the other threads with a local copy of the model and a subset of the dataset. The execution is almost similar to that carried out in algorithm 1, where each slave thread partitions his subset of dataset (d_i) into several batches b_i of size M (line 1). It then goes through each batch b_i, and for each sentence x in b_i, it predicts its class (line 6), compute and accumulates the prediction error as well as the partial derivatives (line 7 and 8). Once a slave thread finishes calculating derivatives, it enters in critical section using a mutex variable to update the global model accessible from shared memory. It then retrieves the updated parameters for the next iterations (line 10 to 12). Finally, it releases the mutex variable (line 13) to allow another thread to perform the update.

4.3 Property

Complexity. Considering algorithm 2 and 3, and let $\hat{I}$ be the index of the subset of data having the largest size $|d_{\hat{I}}|$. The worst-case time complexity for a standard RNN is $O(|d_{\hat{I}}|n_e(n_o n_h + n_s n_h(n_h + n_e))$ and for LSTM and GRU is $O(|d_{\hat{I}}|n_e(n_s n_h + n_s(n_e + n_h(n_h + n_e))))$.

Maximum Speedup. Considering the same maximum epoch number n_e used for the sequential version, the maximum speedup for this parallel implementation corresponds to the number of units of treatment (P) used. In practice it is difficult to have a speedup very close to the maximum speedup, this is mainly due to the portions of code executing sequentially for the central thread.

Convergence. The convergence of the model when trained with this parallel implementation is close to that observed with a sequential implementation. In fact, with only one thread, it is easy to see that parallel algorithm and sequential algorithm have the same convergence. With many slave threads, if we align the execution of each thread one after the other, we can see that the convergence is close to that of sequential execution (as in Algorithm 1). This convergence can be show formally. We plan to show it as it was done in [6] for the parallelization of a sparse SVM.

5 Experiments and Results

In this section we present the experimental results obtained for sequential execution and after for the parallel execution. The experiments were carried out on a multi-core machine with 8 cores (at $2.20\,\mathrm{GHz}$) and 8 Go of Ram. The implementations of the training algorithms were done in C language and using the posix thread library for parallel implementations. We used two datasets. The first dataset (dataset 1) is the one we obtained by performing scraping on facebook pages of Cameroonian users. This dataset contains 3110 non-hate comments and 3069 hate comments. The second dataset, (dataset 2) obtained from Kaggle[3] contains a set of English-language tweets. After sub-sampling it we got 5190 tweets including 1880 offensive tweets, 1430 hateful tweets and 1880 neither offensive nor hateful tweets. The two datasets were split into two, namely 80% for training, and 20% for testing and validation.

Since we are performing a text classification task and the datasets used are balanced, the appropriate metrics performance to evaluate the model are the unweighted averages of: precision, recall, and f-measure. The description and formula of these metrics are given in the Table 1, where l is the number of class, tp_i (true positives), the number of correctly recognized examples for class i, tn_i (true negatives), the number of correctly recognized examples that do not belong to class i, fp_i (false positives), numbers of examples that either were incorrectly assigned to the class i and fn_i (false negatives) the numbers of examples were not recognized for class i [8].

---

[3] https://www.kaggle.com/datasets/mrmorj/hate-speech-and-offensive-language-dataset.

Table 1. Metric measurement descriptions and formula to evaluate the model.

Measure	Description	Formula
Recall	An average per-class effectiveness of a classifier to identify class labels	$\frac{\sum_{i=1}^{l} \frac{tp_i}{tp_i + fp_i}}{l}$
Precision	An average per-class agreement of the data class labels with those of a classifiers	$\frac{\sum_{i=1}^{l} \frac{tp_i}{tp_i + fn_i}}{l}$
F-measure	Relations between data's positive labels and those given by a classifier based on a per-class average	$\frac{2 \times Recall \times Precision}{Recall + Precision}$

The performance due to parallelization of an algorithm will be measured using the following metrcic :

- **Execution time T(p):** Is the execution time taken by a parallel program using p resources.
- **The Speedup S(p):** Is the ratio between the execution time with one resource, on the execution time on p resources.

$$S(p) = \frac{T(1)}{T(p)} \tag{6}$$

5.1 Sequential Execution

The hyperparameter values used for running the learning algorithm for each of the RNN variants are as follows: the maximum number of epochs (15), the learning rate λ (0.1), the batch size M (32), the total number of n_h neurons in the hidden layer (80), the size n_v of input vectors representing a word (30). Table 2 presents the results obtained for the two datasets as well as the training time taken for the training of the model with each of the variants of the RNN.

Table 2. Metric measurements and time taken for each RNN variants.

	Precision	Recall	F-measure	Training time (s)
LSTM$_{\text{dataset2}}$	**0.728**	**0.715**	**0.716**	**1970**
GRU$_{\text{dataset2}}$	0.718	0.710	0.702	1555
RnnStandard$_{\text{dataset2}}$	0.605	0.567	0.549	540
LSTM$_{\text{dataset1}}$	**0.528**	**0.533**	**0.530**	**4361**
GRU$_{\text{dataset1}}$	0.528	0.525	0.511	3134
RnnStandard$_{\text{dataset1}}$	0.239	0.500	0.354	1065

Looking at Table 2, we see that each time we obtain models with better performance when using an LSTM with one or other of the datasets. However, using an LSTM involves a higher training time compared to other RNN variants.

We then implemented the parallel training algorithm of an LSTM hoping to reduce the execution time while keeping the metric measures. We also note that the model obtained with dataset 1 is not efficient enough. It does not allow the neural network to find a correlation between comments classified as hateful or non-hateful. The cause would be the data collection and labeling process that was not carried out optimally.

5.2   Parallel Execution

In order to perform comparisons, we used two parallel implementation strategies for the training algorithm of an LSTM with the dataset 2. The first (parallel without aggregation) is the idea of parallelization that we propose in this paper. The second (parallel with aggregation) is the one carrying out an aggregation (in this case the arithmetic mean) of the various local models to have the final global model. We notice that the saving in time and the speedup is almost the same for each of the two strategies. Using the eight available cores, we obtain a speedup of 2.21 for the strategy without aggregation, and a speedup of 2.19 for the strategy with aggregation. Note that for the parallelization strategy with aggregation, it is possible to use an aggregation function that is more complex than the arithmetic mean, for example to guarantee better convergence, while taking the risk that the gain in time is less important.

We were also interested in the evolution of the convergence of the model for each of the two strategies using graphs showing the evolution of the training error as a function of the number of epochs (see Fig. 2). By examining these graphs, we note that the model converges better with the strategy of parallelization without aggregation compared to that carrying out the arithmetic average of the various local models.

We also note that the number of processing units used has a more negative impact on the strategy with aggregation. We can for example see that with eight processing units, the convergence of the model using this strategy is much worse than that with two processing units. Table 3 presents the performance metrics obtained with each of the two parallelization strategies using the eight processing units available, and in comparison with those obtained with sequential execution. By observing Table 3, we see that for one or the other of the RNNs, we obtain models with better metric performances when we use the parallel implementation strategy using mutex synchronization. We can conclude on the fact that, it would be more interesting to use a parallelization strategy without aggregation compared to that carrying out an aggregation between the local models of the various processing units. One of the main difficulties or disadvantages of strategies using an aggregation is the fact of having to choose the appropriate method or aggregation function guaranteeing good convergence and saving time.

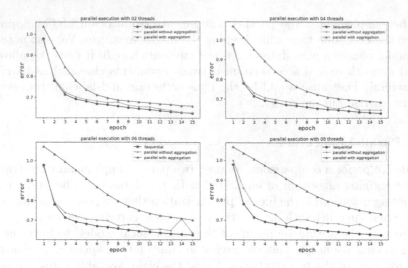

Fig. 2. Error evolution with LSTM using the **dataset 2**

Table 3. Table presenting the precision metrics for parallel implementations (with 08 threads) as well as the time taken to train the model using the **dataset 2**

		Precision	Recall	F-measure	Training time (s)
LSTM	sequential	0.728	0.715	0.716	1970
	parallel without aggregation	**0.726**	**0.699**	**0.700**	**889**
	parallel with aggregation	0.690	0.654	0.651	897
GRU	sequential	0.718	0.710	0.702	1555
	parallel without aggregation	**0.686**	**0.677**	**0.678**	**722**
	parallel with aggregation	0.668	0.652	0.651	730
Standard	sequential	0.605	0.567	0.549	540
	parallel without aggregation	**0.616**	**0.549**	**0.527**	**230**
	parallel with aggregation	0.587	0.522	0.504	227

6   Conclusion

In this paper we first compared sequential executions of standard RNN, LSTM and GRU on two datasets: a dataset gotten by performing scraping on facebook pages of Cameroonian users, and another dataset obtained on Kaggle. We have shown that the LSTM provides better metric performance (an f-measure of 0.71 with an LSTM compared to 0.70 and 0.65 respectively with a GRU and standard RNN on Kaggle dataset). But an LSTM implies a more important execution time (1970 s) in comparison with a GRU (1555 s) and a standard RNN (540 s). We then show that with parallelization, we can get both good metric performance and good execution time. For this issue, we proceeded to a parallel implementation of the training algorithms. We proposed the parallelization strategy with implicit aggregation in comparison to the existing approach which is based on a strategy with an explicit aggregation function. The experimental results on an 8-core machine at 2.20 GHz show that better results are obtained with the parallelization strategy that we proposed. In fact, for the parallel implementation of an LSTM using Kaggle dataset, we got an f-measure of 0.70 and a speedup of 2.2 with an implicit aggregation, compared to a f-measure of 0.65 and a speedup of 2.19 with an explicit aggregation.

As future work, we plan to continue our experiments by using more hardware resources and larger datasets. It would also be interesting to use and compare other deep learning algorithms like transformers, which are successful at the moment especially for Natural Language Processing tasks.

Aknowlegment. The successful completion of this research endeavor has been made possible through the generous support and collaboration of various scholarship stakeholders. We express our sincere gratitude to International Development Research Centre (IDRC), Swedish International Development Cooperation Agency (SIDA) and African Center for Technology Studies (ACTS) for awarding the Artificial Intelligence for Development (AI4D) scholarship programme that funded this research. This scholarship not only provided financial support but also served as a source of motivation and encouragement throughout the project. We are grateful for the support of all members of our work team namely High performance data science (HIPERDAS), who contributed to stimulating discussions and shared their insights. The collaborative environment fostered by our academic community has been integral to the development of this work. Lastly, we extend our thanks to all those who, directly or indirectly, played a role in the realization of this paper. Your support has been crucial, and we are grateful for the opportunities provided by the scholarship stakeholders and the broader academic community.

References

1. Dean, J.: Large scale distributed deep networks. In: Advances in Neural Information Processing Systems, vol. 25 (2012)
2. Schmidt, R.M.: Recurrent neural networks (RNNs): a gentle introduction and overview. arXiv preprint arXiv:1912.05911 (2019)
3. Pascanu, R.: On the difficulty of training recurrent neural networks. In: International Conference on Machine Learning, PMLR (2013)
4. Abadi, M.: Tensorflow: large-scale machine learning on heterogeneous distributed systems. arXiv preprint arXiv:1603.04467 (2016)
5. Huang, Z.: Accelerating recurrent neural network training via two stage classes and parallelization. In: 2013 IEEE Workshop on Automatic Speech Recognition and Understanding. IEEE (2013)
6. Recht, B.: Hogwild!: a lock-free approach to parallelizing stochastic gradient descent. In: Advances in Neural Information Processing Systems, vol. 24 (2011)
7. Chen, G.: A gentle tutorial of recurrent neural network with error backpropagation. arXiv preprint arXiv:1610.02583 (2016)
8. Sokolova, M.: A systematic analysis of performance measures for classification tasks. Inf. Process. Manag. **45**(4), 427–437 (2009)

Two High Capacity Text Steganography Schemes Based on Color Coding

Juvet Karnel Sadié[1,2,3], Leonel Moyou Metcheka[1,2,3],
and René Ndoundam[1,2,3(✉)] (iD)

[1] Team GRIMCAPE, Yaounde, Cameroon
[2] Sorbonne University, IRD, UMMISCO, 93143 Bondy 10587 , France
[3] Department of Computer Science, University of Yaounde I, P.o. Box 812,
Yaounde, Cameroon
ndoundam@yahoo.com

Abstract. Text steganography is a mechanism of hiding secret text message inside another text as a covering message. In this paper, we propose a text steganographic scheme based on color coding. This includes two different methods: the first based on permutation, and the second based on numeration systems. Given a secret message and a cover text, the proposed schemes embed the secret message in the cover text by making it colored. The stego-text is then send to the receiver by mail. After experiments, the results obtained show that our models performs a better hiding process in terms of hiding capacity as compared to the scheme of Aruna Malik et al. on which our idea is based.

Keywords: Steganography · Permutation · Embedding Capacity · Numeration Systems

1 Introduction

The word steganography is of Greek origin and means covered writing. It is the hiding of a message within another (cover medium) such as web pages, images or text, so that the presence of the hidden message is indiscernible. The key concept behind steganography is that the message to be transmitted should not be detectable with bare eyes. From the definition, steganography is used to ensure data confidentiality, like encryption. However, the main difference between the two methods is that with encryption, anybody can see that both parties are communicating in secret. Steganography hides the existence of a secret message and in the best case nobody can detect the presence of the message. When combined, steganography and encryption can provide more security. A number of steganographic methods have been introduced on different cover media such as images [1], video files [2] and audio files [3]. In text based steganographic

P. Melatagia Yonta et al. (Eds.): CRI 2023, CCIS 2085, pp. 166–179, 2024.
https://doi.org/10.1007/978-3-031-63110-8_14

methods, text is used as a cover media for hiding the secret data. Due to the lack of large scale redundancy of information in a text file, the human eye is very susceptible to any change between the original and the modified texts. Therefore, text steganography seems to be the most difficult kind of steganography [4].

Aruna Malik et al. [5] proposed a high capacity text steganography scheme based on LZW compression and color coding. Their scheme uses the forward mail platform to hide secret data. The algorithm first compresses secret data and then hide the compressed data into the email addresses and also, in the cover message of email. The secret data is embedded in the message by making it colored using a color table.

In this paper, we first present some limits of the scheme of Aruna Malik et al. [5] and then propose a text steganographic scheme based on color coding, permutation and numeration systems, which improve the embedding capacity of the scheme of Aruna et al. [5]. Given a secret message and a cover text, the proposed scheme embed the secret message in the cover text by making it colored, using a permutation algorithm for the first method and numeration systems for the second one.

Section 2 presents some preliminaries and related works. Section 3 concerns the presentation of the first method of our scheme. Section 4 labels the second approach of our scheme, Sect. 5 presents experimental results. Discussion and conclusion are stated in Sects. 6 and 7 respectively.

2 Preliminaries and Related Works

In this section, the focus is to present some preliminaries that lead us to the comprehension of our scheme.

2.1 Text Steganography

Several works have been proposed in the field of text steganography [5–9]. Ekodeck and Ndoundam [7] proposed different approaches of PDF file based steganography, essentially based on the Chinese Remainder Theorem. Here, after a cover PDF document has been released from unnecessary characters of ASCII code A0, a secret message is hidden in it using one of the proposed approaches, making it invisible to common PDF readers, and the file is then transmitted through a non-secure communication channel.

Rajeev et al. [8] proposed an email-based steganography method using a combination of compression. The method uses the email forwarding platform to hide secret data in email addresses and the combination of BWT, MTF and LZW compression algorithms to increase the embedding capacity. Rajeev et al. [9] proposed a high-capacity email-based text steganography method using Huffman compression. This method uses the message forwarding platform to hide secret data in email addresses. To increase the embedding capacity, the number of characters in the email address is used to reference the secret bits.

Additionally, the method adds random characters just before the @ symbol to increase randomness.

Aruna Malik et al. [5] proposed a steganographic scheme based on LZW compression and coloring. Their scheme uses the email transfer platform to hide the secret. The algorithm first compresses the secret, then hides this compressed secret in the email addresses and in the body of the email. This concealment proceeds by coloring the text using a color coding table. The Table 1 gives the embedding capacity (CE) of the steganographic schemes presented above.

Table 1. presentation of the embedding capacity (CE) of some steganographic techniques based on coloring and mail transfer

Techniques	CE in percentage
Desoky [10]	3.87 %
Rajeev et al. [8]	7.03 %
Rajeev et al. [9]	7.21 %
Aruna et al. [5]	13.43%

we present below some limits of the scheme of Aruna Malik et al. [5].

2.2   Critic and Limits

LZW is a lossless compression technique that performs high compression ratio when the source contains repetition pattern. In the LZW based steganographic scheme propose by Aruna Malik et al. [5], they apply this lossless compression on the secret message to increase the embedding capacity. But in the example proposed, there is no compression. The secret message is: "underlying physiological mechanisms".

With the LZW Algorithm with initial dictionary fixed and known [11], the size of the secret message after compression is 35*9 = 315 bits;

With the LZW algorithm with sharing of the initial dictionary [11], the size of the secret message after compression is 17 bytes + 35 bytes = 52 bytes = 416 bits;

With the Unix compress command[1] the output indicates that there is no compression and the .Z file size is 44 bytes = 352 bits.

The Table 2 shows the comparison in terms of bit between the original text size and the output size after the compression using the three different approaches of LZW implementation: the LZW Algorithm with initial dictionary fixed and known, the LZW algorithm with sharing of the initial dictionary and the Unix compress command.

---

[1] Linux Compress Command Examples for Files and Directory, https://linux.101hacks.com/unix/compress/.

Table 2. Secret message size comparison

Secret message size	Output size 1	Output size 2	Output size 3
280	315	416	352

From Aruna et al. paper [5], the size obtained was 264 bits, but we have proven above that there is no compression for this example. This is the principal limit of this steganographic scheme, where for some messages the reduction of the message size will not be possible.

Furthermore, another limit to the scheme of Aruna et al. paper [5] is that:**for any integer n, there are 2^n different binary words of length n, but only $\sum_{i=0}^{n-1} 2^i = 2^n - 1$ shorter descriptions.** For all n, there therefore exists at least one binary word of length n which cannot be compressed [12]. So there will be cases where the secret cannot be compressed. Our paper uses:

- The idea of color coding contained in the paper of Aruna Malik et al. [5];
- The permutation generation method of W. Myrvold and F. Ruskey [13];
- The numeration systems;

to present a new scheme where the secret message embedding capacity is better than the scheme of Aruna Malik et al. [5].

2.3 Permutation Generation Methods

Permutation is one of the most important combinatorial object in computing, and can be applied in various applications, for example, the scheduling problems. Permutation generation has a long history. Surveys in the field have been published in 1960 by D.H. Lehmer [14]. Several authors [13,15,16] have since developed many methods to generate all the possible permutations of n elements. Also, several works [17,18] in steganography taking advantage of permutations have been done. W. Myrvold and F. Ruskey [13] proposed a ranking function (Algorithm 2) for the permutations on n symbols which assigns a unique integer in the range $[0, n! - 1]$ to each of the $n!$ permutations. Also, they proposed an unranking function (Algorithm 1) for which, given an integer r between 0 and $n!$ - 1, the value of the function is the permutation of rank r.

Unranking Function. First of all, recall that a permutation of order n is an arrangement of n symbols. An array $\pi[0 \cdots n - 1]$ is initialized to the identity permutation $\pi[i] = i$, for $i = 0, 1, \cdots n - 1$.

Algorithm 1 :*Unranking function*

```
Procedure unrank(n, r, π)/13/
    begin
        if n > 0 then
            swap(π[n − 1], π[r mod n]);
            unrank(n − 1, ⌊r/n⌋ , π);
        end;
    end;
```

Note: $swap(a, b)$ exchanges the values of variables a and b.

Ranking Function. To rank, first compute π^{-1}. This can be done by iterating $\pi^{-1}[\pi[i]] = i$, for $i = 0, 1, \cdots, n - 1$.
In the algorithm below, both π and π^{-1} are modified.

Algorithm 2 :*Ranking function*

```
function rank(n, π, π⁻¹):integer[13]
    begin
        if n = 1 then return(0) end;
        s := π[n − 1] ;
        swap(π[n − 1], π[π⁻¹[n − 1]]) ;
        swap(π⁻¹[s], π⁻¹[n − 1]) ;
        return(s + n.rank(n − 1, π, π⁻¹)) ;
    end;
```

3   Scheme Design Based on Permutation

In this section, we present the first method of our scheme. The proposed algorithm takes the cover text C, the secret message M, the e-mail address of the receiver and the initial permutation. First, it computes the binary representation of the secret and divides that representation into blocks of t bits, $t = \lfloor log_2(n!) \rfloor$. It also divides the cover-text into blocks of n characters. For each block of the secret stream, computes its decimal representation (the rank) and the permutation related to that rank, using the initial permutation. Colors the corresponding block of the cover text according the permutation obtained. It finally concatenates all the stego-blocks and send them to the receiver by e-mail.

3.1   Embedding Algorithm

Input:
C: the cover text;
M: the secret message to embed;
The key π: the initial permutation of n colors;

e: the e-mail address of the receiver;

Output:

C': the stego-message;

begin:

1. Compute m, the binary representation of M;
2. Compute $t = \lfloor log_2(n!) \rfloor$
3. Divide m into p blocks of t bits each, $b_1, b_2, \cdots, b_p$;
4. Divide C into k blocks of n characters each $c_1, c_2, \cdots, c_k$;
5. For each block $b_i, 1 \leq i \leq p$:

    a. compute $Nperm = (b_i)_{10}$, the decimal representation of b_i;

    b. compute $\pi' = unrank(n, Nperm, \pi)$, the permutation corresponding to the number $Nperm$. π' can be considered as $\pi'(1), \pi'(2), \cdots, \pi'(n)$;

    c. color each character of c_i by the corresponding color given by the permutation π' and obtain the string c_i';

    d. compute $C' \leftarrow C' || c_i'$; where a||b is the concatenation of a and b.

6. If the next character is EOF (End of File) then

    begin

    a. Use e to send C' by mail to the receiver;

    end

    Else

    begin

    a. Colour the next character with a color different of permutation colors. This color is shared by the sender and the receiver. However, this color will not be very distant from the others;

    b. Randomly color the rest of characters of C by the colors of colors table, until you obtain the EOF character;

    c. Use e to send C' by mail to the receiver;

    end;

  end;

3.2 Retrieval Algorithm

Given the stego-text C', the algorithm extract the secret message M, using the initial permutation. First the stego-text is divided into blocks of n characters. For each block, the algorithm extract the rank related to the permutation according to the order of appearance of the colors. Finally the algorithm concatenates the binary representation of the rank and obtains the secret.

Input:

C': the stego-text;

The key π: the initial permutation of n colors;

Output:

M: the secret message;

begin:

1. Retrieve all characters coloured by the permutation colors, until a color different from the colors in the colors table, or the EOF character is obtained. Lets call them C'';

2. Divide C'' into p blocks of n characters each $c_1, c_2, \cdots, c_p$;
3. For each block $c_k, 1 \le k \le p$:
      a. use the color order of characters to compute the relative permutation, that we call π'. π' can be considered as $\pi'(1), \pi'(2), \cdots, \pi'(n)$;
      b. compute the number $Nperm = rank(n, \pi', \pi'^{-1})$;
      c. compute $m' = (Nperm)_2$, the binary representation of $Nperm$;
      d. compute $M \leftarrow M || m'$;
end;

In order to increase the security of the model, the secret message can first be encrypted by the AES method with a key shared between the two communicating parties. On arrival, after extracting the secret data, the recipient will then have to decipher it to obtain the secret message.

4   Scheme Design Based on Numeration Systems

In this new approach, we improve the method of the first scheme with the assertion that each color can be repeated as many times on some positions of a given group of characters. Unlike the previous scheme in which each color could only appear once in a group of precise characters.

4.1   The Scheme Description

we give a brief description of how this new scheme works by following these steps:

1. Choose a base B such that $2 \le B \le 2^{24}$. where 2^{24} is the number of existing colors ;
2. choose B colors from the set of 2^{24} colors number from 0 to $B - 1$;
3. convert the secret m to base B such that : $m = (m_{q-1}...m_1 m_0)_B$, where $0 \le m_i \le B - 1$;
4. We assume that the number of characters of the covert text is n and $q \le n$;
5. For $i = 0$ to $q - 1$ do
   The character c_i is coloured with the color relative to m_i
6. The text coloured is then send to the receiver.

The reverse procedure consists to extract the secret conceal in the colors distribution. These steps must be performed by the receiver of the stego-text :

1. Take the text with the first q characters which has been coloured;
2. For $i = 0$ to $q - 1$ do

Find the color number z_i associated to the character c_i by using the reference color table shared between the sender and the receiver;
3. Convert $z = (z_{q-1}...z_1 z_0)_B$ to binary and get the secret message.

4.2   Embedding Algorithm

Input
C: the cover text; M: the secret message to embed; β : The base; T : a table of β color; e : the e-mail address of the receiver;
Output
C': the stego-message;
Begin

1. Convert the secret M to base B such that : $m = (m_{n-1}...m_1 m_0)_B$, where $0 \leq m_i \leq B - 1$;
2. For $i = n - 1$ to 0 do
   (a) Find in the color table, the color a_i associated to the value m_i;
   (b) Colour the character c_i of C with the color a_i and obtain c_i' ;
   (c) Compute $C' \longleftarrow C' \parallel c_i'$; where $a \parallel b$ is the concatenation of a and b;
3. If the next character is not EOF (End of File) then
   (a) Colour the next character with a color different from the colors table T. This color is shared by the sender and the receiver. However, this color will not be very distant from the others;
   (b) Randomly color the rest of characters of C by the colors from the colors table, until you obtain the EOF character;
   (c) Compute $C' \longleftarrow C' \parallel c_j' : n \leq j \leq m$, where m is the position of the last character of C;
4. Use e to send C' by mail to the receiver;

End

4.3   Retrieving Algorithm

Input
C': the stego-text; T : a table of β color; β : The base;
Output
M: the secret message;
Begin

1. Retrieve all characters coloured with the table colors, until obtain a color different from those of the colors table, or obtain the EOF character. Lets call them C'' and $\mid C'' \mid = n$; $(C'' = c_{n-1} c_{n-2}...c_1 c_0)$;
2. For $i - n - 1$ to 0 do
   (a) get the color a_i associated to the color of the character c_i of C'';
   (b) Find in the color table, the value m_i associated to the color a_i ;
   (c) compute $M \longleftarrow M \parallel m_i$;
   (d) Compute M_2, the binary representation of the secret M;

End

   As we previously said, the secret message can first be encrypted by the AES method with a key shared between the two communicating parties. On arrival, after extracting the secret data, the recipient will then have to decipher it to obtain the secret message.

5 Experimental Results

In this section, we present some experimentations related to our scheme.

5.1 First Method

In this subsection, we first propose a theoretical estimation of our embedding capacity for n colors. Secondly, we present practical experimentation results in the case of 10, 16, 32 and 64 colors, based on example 1 of [5]. The table of 10 colors is given in Fig. 1.

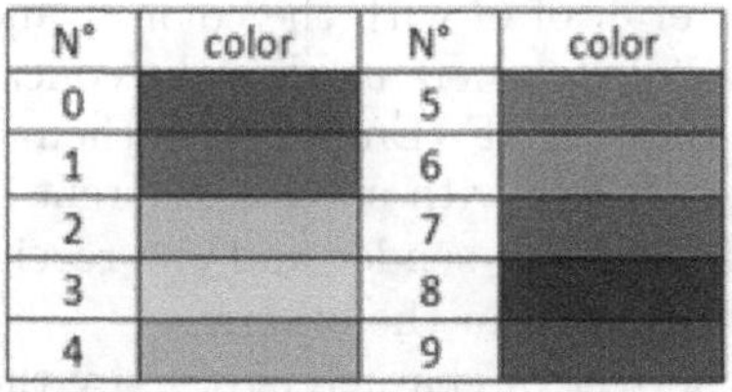

N°	color	N°	color
0		5	
1		6	
2		7	
3		8	
4		9	

Fig. 1. The table of 10 colors

Theoretical Estimations. The Table 3 presents the embedding capacity of our scheme for some different values of n: 10, 16, 32,64. This theoretical estimation is based on our embedding algorithm.

More generally, in a set of n colors, the number of permutation of n distinct colors is n !. According to the stirling formula [19] we have:

$$n! \sim \left(\frac{n}{e}\right)^n \times \sqrt{2\pi n} \tag{1}$$

where $\pi = 3.14$ is the area of the circle with unit radius, $e = 2.718$ is the base of the natural logarithm, and $\sim$ means approximate equality.

we know that:

$$n = 2^{log_2(n)} \tag{2}$$

By replacing the value of n in Eq. 1 we have:

$$n! \sim \left(\frac{2^{log_2(n)}}{2^{1.442695}}\right)^n \times \sqrt{2\pi n} \tag{3}$$

$$\sim \left(2^{log_2(n)-1.442695}\right)^n \times \sqrt{2\pi n} \tag{4}$$

$$\sim \left(2^{nlog_2(n)-1.442695n}\right) \times \sqrt{2\pi n} \tag{5}$$

$$\sim \left(2^{nlog_2(n)-1.442695n}\right) \times 2^{log_2(\sqrt{2\pi n})} \tag{6}$$

$$\sim \left(2^{nlog_2(n)-1.442695n}\right) \times 2^{\frac{1}{2}log_2(2\pi n)} \tag{7}$$

$$\sim \left(2^{nlog_2(n)-1.442695n+\frac{1}{2}log_2(2\pi n)}\right) \tag{8}$$

Proposition : the embedding capacity (E) using n colors to hide a secret is :

$$E = \frac{M \times 100}{n \times 8}$$

where $M = n(log_2(n) - 1.442695) + \frac{1}{2}log_2(2\pi n)$, and n the number of colors.

Table 3. Theoretical estimations of the proposed scheme

n	M= $\lfloor log(n!) \rfloor$	P=M/8	100*(P/n),(embedding capacity)
10	21	2.6	26.25%
16	44	5.5	34.37%
32	117	14.6	45.63%
64	295	36.9	57.66%

Remark: As far as the space characters of the stego-text are not coloured, the embedding capacity can decrease in the experimentations.

Experimentation. Here, the secret message is : **underlying physiological mechanisms**
and the cover text is:
Only boats catch connotes of the islands sober wines only ships wrap the slips on the cleats of twining lines only flags flap in tags with color that assigns only passage on vessels
We apply our embedding algorithm and obtain the stego-text given by the Fig. 2. That stego-text is then send by mail to the receiver.

Only boats catch connotes of the islands sober wines only ships wrap the slips on the cleats of twining lines only flags flap in tags with color that assigns only passage on vessels

Fig. 2. The stego-text

With this example:

– in the case of 10 colors, the embedding capacity is 20.58 %;
– with 16 colors, the embedding capacity is 25.5 %;
– with 32 colors, the embedding capacity is 29.5 %;
– with 32 colors, the embedding capacity is 45.45 %;

5.2   Method 2

Theoretical Estimation. We want to color a block of text with η characters. Each character is coloured with a single color. The number of colors used is B. Knowing that a color can appear as many times on some positions, the total number of colouring possibilities for each character is : B. For the η characters, the total number of colouring possibilities is : B^η. The number of bits used to color the η characters is : $log_2(B^\eta)$. The embedding capacity [20] is define as the ratio of the secret bits message by the stego cover bits :

$$Capacity = \frac{log_2(B^\eta)}{\eta \times 8} = \frac{log_2(B)}{8} \tag{9}$$

The Table 4 gives a theoretical estimation of the capacity as a function of the base B used.

Table 4. Theoretical estimations of the proposed scheme

B	Capacity $\times 100$
10	41.5%
16	50%
32	62.5%
64	75%

Remark: As far as the space characters of the stego-text are not coloured, the embedding capacity can decrease in the experimentations.

Experimentation. This experimentation is based on example 1 of [5], where the number of color B is equal to 10. The Fig. 3 presents the results of the embedding process based on this second method for 10 colors.

Only boats catch connotes of the islands sober
wines only ships wrap the slips on the cleats of
twining lines only flags flap in tags with color
that assigns only passage on vessels

Fig. 3. The stego-text for a table of 10 colors

With this example :

– in the case of 10 colors, the embedding capacity is 34.31 %;
– with 16 colors, the embedding capacity is 41.17 %;
– with 32 colors, the embedding capacity is 52.23 %;

6   Discussion

The aim of this paper is to propose a new scheme that improves the scheme proposed by Aruna et al. [5]. The Table 5 recapitulates the embedding capacity of our schemes in comparison with the scheme of Aruna et al. [5], in the case of 10 colors.

Table 5. Comparison between our scheme and the scheme of Aruna et al. [5], in terms of embedding capacity, for 10 colors

	First Method	Second Method	Aruna et al. [5]
example 1 [5]	**20.58 %**	**34.31 %**	6.03 %
example 2 [5]	**22.32 %**	**35.29 %**	13.43%

More Generally, the Fig. 4 gives the graphical representation of the performance of our scheme compared to other schemes.

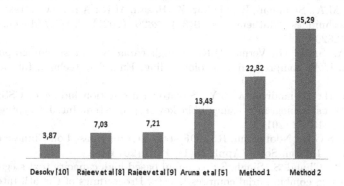

Fig. 4. The graphical representation of the performance of our scheme compared to other schemes

It is true that coloring the text brings attention and then raises suspicion. We think that to solve this problem and provide a better undetectability, the cover text can be classified into the following fields: painting, poetry, art, humor, love. The color set is given by the initial permutation(the key). Before the communication, the two parties first share it by mail for instance.

7   Conclusion

In this paper, two text steganographic schemes based on color coding have been proposed. The first based on permutation and the second based on numeration systems. Given a secret message and a cover text, the proposed schemes embed the secret message in the cover text by making it coloured. These two high capacity text steganographic schemes significantly improve the existing work of Aruna et al.

Acknowledgments. This work was supported by *UMMISCO, LIRIMA*, and the *University of Yaounde 1*. The authors are grateful for this support.

Disclosure of Interests. The authors have no competing interests.

References

1. Subramanian, N., Elharrouss, O., Al-Maadeed, S., Bouridane, A.: Image steganography: a review of the recent advances. IEEE Access **9**, 23409–23423 (2021). https://doi.org/10.1109/ACCESS.2021.3053998
2. Kunhoth, J., Subramanian, N.: Somaya Al-ma adeed, Ahmed Bouridane, "Video steganography: recent advances and challenges." Multimed. Tools Appl. **82**(27), 1–43 (2023). https://doi.org/10.1007/s11042-023-14844-w
3. Tanwar, R., Bisla, M.: Audio steganography. In: 2014 International Conference on Reliability Optimization and Information Technology (ICROIT), Faridabad, India, 2014, pp. 322–325 (2014). https://doi.org/10.1109/ICROIT.2014.6798347.
4. Majeed, M.A., Sulaiman, R., Shukur, Z., Hasan, M.K.: A review on text steganography techniques. Mathematics. **9**(21), 2829 (2021). https://doi.org/10.3390/math9212829
5. Malik, A., Sikka, G., Verma, H.K.: A high capacity text steganography scheme based on LZW compression and color coding. Eng. Sci. Technol. Int. J. **20**, 7127 (2017)
6. Kabetta, H., Dwiandiyanta, B.Y., Suyoto.: Information hiding in CSS: a secure scheme text steganography using public key cryptosystem. Int. J. Cryptogr. Inform. Security **1**, 13–22 (2011)
7. Ekodeck, S.G.R., Ndoundam, R.: PDF steganography based on Chinese remainder theorem. J. Inform. Secur. Appl. **29**(1), 1–15 (2016)
8. Kumar, R., Chand, S., Singh, S.: An email based high capacity text steganography scheme using combinatorial compression. In: Proceedings of the 5th International Conference - Confluence The Next Generation Information Technology Summit, pp. 336–339 (2014)
9. Kumar, R., Malik, A., Singh, S., Chand, S.: A high capacity Email based text steganography scheme using Huffman compression. In: Proceedings of the International Conference on Signal Processing and Integrated Networks (2016)
10. Desoky, A.: list-based steganography methodology. Int. J. Inform. Secur. **8**, 247–261 (2009)
11. Wang, Z.H., Yang, H.R., Cheng, T.F., Chang, C.C.: A high-performance reversible data-hiding scheme for LZW codes. J. Syst. Softw. **86**(11), 2771–2778 (2013)
12. Martin, B.: Codage, cryptologie et Applications. In: Presses Polytechniques et Universitaires Romanes, p. 32
13. Myrvold, W., Ruskey, F.: Ranking and unranking permutations in linear time. Inf. Process. Lett. **79**(6), 281–284 (2001)
14. Lehmer, D.H.: Teaching combinatorial tricks to a computer. In: Proceedings of Symposium in Applied Mathematics, Combinatorial Analysis, American Mathematical Society, Vol. 10, pp. 179–193 (1960)

15. Nijenhuis, A., Wilf, H.S.: Combinatorial Algorithms: For Computers and Calculators, 2nd Edition. Academic Press, New York (1978)
16. Djamégni, C.T., Tchuente, M.: A cost-optimal pipeline algorithm for permutation generation in lexicographic order. J. Parall. Distrib. Comput. **44**(2), 153–159 (1997)
17. Al-Bahadili, H.: A secure block permutation image steganography algorithm. Int. J. Cryptogr. Inform. Secur. **3**(3), 11–22 (2013)
18. Sadié, J.K., Ekodeck, S.G.R., Ndoundam, R.: Binary image steganography based on permutation. Iran J. Comput. Sci. https://doi.org/10.1007/s42044-023-00142-z (2023)
19. Leversha, G., Lovacz, L., Pelikan, J., Vesztergombi, K.: Discrete mathematics, elementary and beyond. The Mathematical Gazette (Springer), Jul 88(512), 378–379, pp. 290 (2004)
20. Satir, H., Isik, A.: compression-based text steganography method J. Syst. Softw. **85** (2012)

Integration of the Triple Block Data Security Model Based on Distributed Crypto-Steganography in a Cluster

Derrick Méthode Bagaza[1] , Blaise Omer Yenke[2(✉)] ,
and Jean M'boliguipa[1]

[1] Faculty of Science, University of Bangui, Bangui, Central African Republic
[2] Department of Computer Science, University Institute of Technology,
University of Ngaoundere, Ngaoundere, Cameroon
boyenke@univ-ndere.cm

Abstract. Cloud environments offer many services to users according to their requests. One of the main concerns of users is to guarantee the availability and integrity of their stored data. Many solutions have been proposed to further ensure the security of this data. This paper is based on one of the recently published solutions that presented a triple-block data security model based on distributed crypto-steganography. The present work shows how to integrate the published security model in any type of distributed system such as Cloud, Grids or Clusters Computing. The integration of this solution is tested in a real cluster environment and data integrity tests showed the effectiveness of the deployment.

Keywords: Distributed Systems · Data Security · Crypto-Steganography

1 Introduction

Cloud Computing is one of today's most powerful technologies, simplifying data storage and disaster recovery. It enables businesses to make more efficient use of their investment in hardware and software as have said Boss et al. (2007) [2]. It also provides users with efficient and more reliable IT services, such as email, instant messaging and web services, at lower cost according to Khalil et al. (2014) [6].

Cloud Computing still does not have a commonly accepted definition, as indicated by Wang and Mu (2011) [12]. According to Rosalina (2020) [10], the National Institute of Standards and Technology (NIST) has defined five essential characteristics of the Cloud, namely: on-demand self-service, broad network access, resource pooling, elasticity or rapid expansion and measured service.

Kartit et al. (2016) [5] indicated that Cloud computing architecture consists of three layers: software as a service (SaaS), platform as a service (PaaS) and infrastructure as a service (IaaS). Islam et al. (2019) [4] have indicated that

P. Melatagia Yonta et al. (Eds.): CRI 2023, CCIS 2085, pp. 180–190, 2024.
https://doi.org/10.1007/978-3-031-63110-8_15

Cloud Computing has for (04) deployment models, namely: the Public Cloud in which the physical infrastructure is managed by the service provider, the Community Cloud in which the physical infrastructure is managed by a group of organisations, the Private Cloud in which the infrastructure is managed by a separate organisation, and the Hybrid Cloud which includes the fusion of the three previous models.

Despite the enormous benefits that Cloud Computing brings to individuals and organisations, there are a number of security issues that require attention in terms of data protection, network security, virtualisation security, program integrity and user management. As a result, a number of mechanisms such as cryptography, steganography and digital signatures have been employed to overcome the problems presented by Mell et al. (2011) [7].

This paper presents the general architecture of the integration of the published triple-block data security model based on distributed crypto-steganography, in any type of distributed system such as a cluster for exemple. The remainder of this article is organized as follows: Sect. 2 is devoted to the generality of distributed systems and clusters. Section 3 describes the proposed deployment architecture and Sect. 4 is dedicated to results and discussions. The conclusion and future work are presented in Sect. 5.

2 General Information on Distributed Systems and Computing Clusters

This section presents the state of the art in distributed systems and Cloud Computing, focusing on their definitions, characteristics, deployment models, advantages and areas of application.

2.1 Distributed Systems

According to Veríssimo et al. (2001) [11], a distributed system is one that prevents you from working when a machine you have never heard of breaks down. According to Ozsu and Valduriez (1994) [9], a distributed system is a number of autonomous (not necessarily homogeneous) processing elements that are interconnected by a computer and cooperate in performing the tasks assigned to them. Generally speaking, a distributed system can be defined as a computer system in which the machines are autonomous and linked together by a network for the purpose of exchanging information. A distributed system is characterised by transparency, reliability, availability, scalability and fault tolerance. However, it offers enormous advantages such as scalability, better performance than a central system, lower cost than a computer and reliability.

Currently, distributed systems are taking a preferential place in almost all areas of life according to the work of Chiluka et al. (2015) [3,8]. Distributed systems can be used in on-demand computing services, high-performance computing, social networks, search engines, banks, online sales sites and in the Cloud.

2.2 Computing Clusters

Network. It is made up of two (02) or more computers, interconnected by a local, often high-speed, network. Each device in a cluster is called a compute node and has one or more compute units and a local memory. All the compute nodes in a cluster work together as a single parallel device. In general, a computing cluster has a node, known as the front-end, whose role is to manage resources and distribute computations across the nodes. According to Ayachi (2019) [1], the nodes in a cluster can be linked together by several communication roles with high bandwidth, such as Gigabit Ethernet, InfiniBand and Myrinet.

According to Yenke et al. (2008) [13], Clusters offer a very good performance/cost ratio and a large aggregate memory capacity, making them a serious competitor to parallel machines.

The following section presents the deployment model based on one of the recently published solutions, presenting a triple block data security model based on distributed crypto-steganography.

3 Proposed Deployment Model

In this section, we propose a model for deployment in a distributed computing cluster. Firstly, the triple-block data security model based on distributed crypto-steganography in the Cloud is presented. Secondly, the cluster deployment model entitled LSBDCT, developed for the purpose of ensuring the security of data stored in Cloud environments, is comprehensively presented.

3.1 Data Security

Information is protected by cryptography and steganography, the two (02) main data security techniques used in online storage environments. This section presents the mathematical tools used by these techniques to secure information.

Cryptographic Technique. This is a technique for transforming secret data from a readable to an unreadable format. It is also the art of securing sensitive data. It has many terminologies, such as clear file, encrypted file, encryption and decryption. The clear text or file is the secret data that the owner wants to make unreadable. Before it is sent, it is converted into another format that is incomprehensible to unauthorised users and is called an encrypted file [14].

Steganographic Technique. Steganography is the technique or art of hiding data in media in order to transmit messages between a sender and a recipient. However, there are three categories of steganography: secret key steganography, public key steganography and pure steganography. Its advantage over cryptography is that the secret message reserved does not attract the attention of a third party.

Mathematical Reminder. The proposed deployment model is based on the concepts of the least significant bit and the discrete cosine transform. For a given medium (image) of size $M \times N$, the DCT is calculated by the following equation:

$$DCT_{kl} = a_k a_l \sum_{m=0}^{M-1} \sum_{n=0}^{N-1} \cos\frac{\pi(2m+1)k}{2M} \cos\frac{\pi(2m+1)l}{2N} \tag{1}$$

$$0 \leq k \leq M - 1 \ et \ 0 \leq l \leq N - 1$$

$$a_k = \begin{cases} \frac{1}{\sqrt{M}} \ for \ k = 0 \\ \sqrt{\frac{2}{M}} \ for \ 1 \leq k \leq M - 1 \end{cases}$$

$$a_l = \begin{cases} \frac{1}{\sqrt{N}} \ for \ l = 0 \\ \sqrt{\frac{2}{N}} \ for \ 1 \leq l \leq N - 1 \end{cases}$$

where DCT_{kl} are the DCT coefficients of row k and column l and C are the values of the pixels of the original image in row m and column n.

This mathematical formula has been integreted into LSB-DCT proposed by Bagaza et al. [14].

3.2   Data Security Model in the Cloud

Hybrid cryptography is used to encrypt secret data using AES and RSA algorithms to achieve a higher level of security. In fact, it takes advantage of the interests of both symmetric and asymmetric encryption, such as the speed boost of symmetric encryption and the security of key sharing by asymmetric encryption.

Distributed steganography is therefore applied to encrypted data that is merged into the pixels of an image in order to protect it in a security-enhancing cover object using least significant bit (LSB) and discrete cosine transform (DCT) techniques. This combined approach of the two techniques aims to transmit data in such a way that no one can detect the existence of the data between a transmitter and a receiver.

We would like to remind you that at present, the cloud is attracting a lot of interest in data security, network security, virtualisation security, application integrity and identity management. As a result, several works have attempted to solve all these problems by using mechanisms such as cryptography, digital signature and steganography to solve these problems. However, the latter reveal some limitations presented above. To this end, the model proposed in this work will further enhance the security of data in the cloud. It is organized in three blocks, in which each of the blocks performs its function properly to ensure data security.

The first block is responsible for authenticating the user. The second block is responsible for the encryption and integration of the secret data into a cover object (media or image) for backup.

The third block is responsible for the recovery and decryption of the secret data. With respect to this organisation, user authentication is used to verify that the data is not tampered with. Once the user authenticates, he or she can perform encryption and data integration operations.

If a malicious entity gains access to the system using illegal means, encryption and data integration using cryptographic and steganographic algorithms can provide an additional layer of security.

In this block, the data is encrypted by AES and RSA algorithms and then the resulting encryption is embedded in an image using the LSB-DCT technique, even if the user credentials have been used fraudulently, thanks to this block, the malicious entity will still not be able to access the secret data in the Cloud. Finally, data recovery, thanks to the LSB-DCT extraction algorithm, decryption by AES and RSA algorithms. This block allows the complete restoration of the data in case of damage.

Figure 1 shows the block diagram of the proposed model. It shows that the approach provides three block security for data sharing and storage in the cloud.

This section presents the triple-block data security model proposed by [14] based on distributed crypto-steganography in the Cloud environment.

The deployment model presented in this work is based on the work of Bagaza et al. (2023) [14] in order to make an effective integration in a real cluster environment.

3.3   Presentation of the Deployment Model

The deployment model is structured in three blocks (b1, b2, b3), in which each of the blocks correctly performs its role in guaranteeing data security. The first block is responsible for encrypting and decrypting user data.

The second block is responsible for generating the different keys for the secret data transfer, for storage in the database. The third block stores the data chirped by block 1 in a distributed fashion. Figure 2 illustrates the proposed cluster deployment model.

3.4   Operating Principle

Any user wishing to submit a task to the cluster must log in as a user in order to benefit fully from the cluster's services. Once logged in, the user can choose between two (02) operations: secure data storage and secure data retrieval.

The data storage operation involves data encryption. To perform encryption, the user must provide the LSB-DCT server with three (03) parameters (operation number, private key and file) to the program before it can be executed.

The program works on the command line to store data from multiple users.

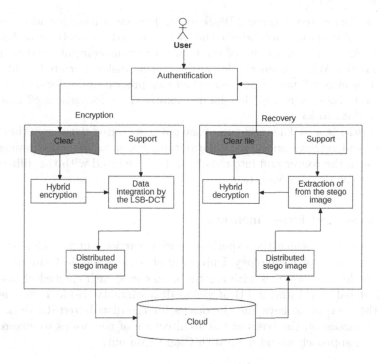

Fig. 1. Proposed architecture.

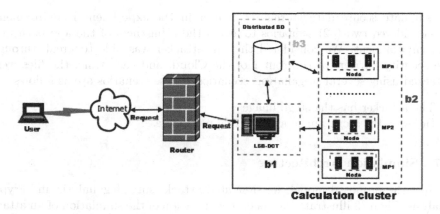

Fig. 2. Overall deployment architecture.

- The block ($b1$): this is the central block or front end, where the basic LSB-DCT encryption and decryption algorithms are deployed. When data arrives from the router, LSB-DCT sends the information in a distributed fashion to the various nodes in the block, to be encrypted or decrypted and then sent to a distributed database. Block $b1$ implements a resource access policy, requiring all users to authenticate themselves in order to deter attackers.

- Block 2 ($b2$): this is made up of the Cloud's various compute and temporal storage nodes. When an encryption or decryption task is initiated by $b1$, these tasks are sent to $b2$ machines. These machines process and inform $b1$ to make the final decision. When $b1$ decision is to store the information, $b2$ machines send the data to $b3$.
- Block 3 ($b3$): is a SGBD made up of several distributed databases that serve to store and retrieve encrypted information. So, if an attacker manages to get through the router and into the cloud, he'll be faced with the difficulty of obtaining the data in plain text.

3.5   Analysis and Experimentation

The environment in which the experiments were carried out is a cluster of the University Institute of Technology, University of Ngaoundere in Cameroon. The cluster is made up of 4 nodes with a total of 56 cores, each clocked at 2.4 GHz, 38 GB of global RAM and 2 TB of global disk capacity. Node 1 was used to generate the encryption keys and the others for the distributed database. This cluster was chosen on the basis of the architecture of the nodes to ensure that the proposed approach would fit in with Cloud Computing.

3.6   Experimental Scenarios

To simulate secure data storage by a user in the experimental environment, we considered two (02) scenarios to assess the robustness of the approach in a real environment. We considered that an attacker was able to break through the security barriers of the router of the Cloud, and each time, the files were attacked using several steganalysis scenarios. These scenarios are as follows:

- The attacker has the stego image
- He owns the stego image and the original image.

3.7   Simulation Architecture

To test the proposed model, we simulated attacks sing steganalysis and cryptanalysis. Figure 3 illustrate the general oganisation of the simulation of an attack on the system.

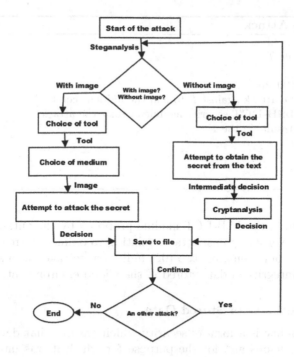

Fig. 3. Attack simulation architecture.

3.8 Running the Scenarios

With the layout of the various elements of the deployment architecture presented above, the application code is deployed on the front-end. Any user wishing to submit an activity to the cluster must access user mode through his account in order to benefit fully from its services.

Users can choose between two (02) operations, namely data encryption and data recovery. Depending on the operation selected, the front-end performs the tasks requested. The following section presents and analyzes the results obtained.

4   Results and Discussion

In the remainder of this work, we refer to *NTADS* as the number of successful attempts to access the stored data, *NEADS* as the number of failed attempts to access the stored data and nt as the number of attempts.

During the tests carried out, quantities were collected and the data saved in trace files for analysis and discussion. Algorithm 1 is used when the scenarios are run and is presented as follows:

Algorithm 1. Attack

Require: F stego file
Ensure: Trace file T
 1: $nt \leftarrow 1$
 2: **while** $nt \leq 100$ **do**
 3:     Launch the attack against F to find the hidden secret
 4:     Run the **LSB-DCT** security module against the attack
 5:     Save the results in the T
 6:     $nt \leftarrow nt + 1$
 7: **end while**

 8: Return T

In Algorithm 1, the LSB-DCT module proposed by the authors [14] is used to ensure data security against attacks relating to online information theft. For the execution of our scenarios, two (02) indices were taken into account: confidentiality and integrity of data stored in the Cloud environment.

4.1 Confidentiality of Stored Data

Data confidentiality is a form of security which ensures that data is used only by authorized persons and for the purpose for which it was intended. Table 1 shows experiments carried out with parameters $nt = 100$, $NTADS = 0$ and $NEADS = 100$.

Table 1. Number of system access attempts.

nt	Scenario 1		Scenario 2	
	$NTADS$	$NEADS$	$NTADS$	$NEADS$
1	No	Yes	No	Yes
2	No	Yes	No	Yes
3	No	Yes	No	Yes
4	No	Yes	No	Yes
5	No	Yes	No	Yes
.	.	.	.	.
.	.	.	.	.
.	.	.	.	.
100	No	Yes	No	Yes

It is clear that the system is 100% secure as measured by the number of successful attempts to access the stored data ($NEADS$) against 0%$NTADS$. Table 1 shows the effectiveness of the system in ensuring the confidentiality of data stored in Cloud environments. The following section presents the effectiveness of the system in ensuring data integrity in Cloud Computing.

4.2 Integrity of Stored Data

In a computer system, the integrity of the data is paramount in order to guarantee the validity of the data and its analysis. From the data in the trace files relating to the attacker's malicious activities on stored data, it emerges that the attacker is able to access the data but cannot do anything to it because he cannot alter or delete it.

In view of the above, the model is 99% correct and 1% is for risk and error. In this way, the system guarantees the integrity of the data in Cloud Computing.

In view of the results presented in the Table 1 and previous section, the model proposed by [14] does indeed prove robust against attacks linked to online data theft. This effectiveness benefits from the advantages provided by the hybridity (combination) of cryptography and steganography.

5 Conclusion

This article is based on one of the recently published solutions presenting a triple-block data security model based on distributed cryptosystems. It presents the integration model of this solution in a real cluster computing environment. Data integrity tests showed the effectiveness of the deployment. Consequently, the proposed model guarantees data security in Cloud environment.

References

1. Ayachi, M.: Etude de la distribution de calculs creux sur une grappe multi-coeurs (2019)
2. Boss, G., Malladi, P., Quan, D., Legregni, L., Hall, H.: Cloud computing. IBM White Paper **321**, 224–231 (2007)
3. Chiluka, N., Andrade, N., Pouwelse, J., Sips, H.: Social networks meet distributed systems: towards a robust sybil defense under churn. In: Proceedings of the 10th ACM Symposium on Information, Computer and Communications Security, pp. 507–518 (2015)
4. Islam, S.J., Chaudhury, Z.H., Islam, S.: A simple and secured cryptography system of cloud computing. In: 2019 IEEE Canadian Conference of Electrical and Computer Engineering (CCECE), pp. 1–3. IEEE (2019)
5. Kartit, Z., et al.: Applying encryption algorithm for data security in cloud storage. In: Advances in Ubiquitous Networking. LNEE, vol. 366, pp. 141 154. Springer, Singapore (2016). https://doi.org/10.1007/978-981-287-990-5_12
6. Khalil, I.M., Khreishah, A., Azeem, M.: Cloud computing security: a survey. Computers **3**(1), 1–35 (2014)
7. Mell, P., Grance, T., et al.: The nist definition of cloud computing (2011)
8. Noumowe, R.C.N.C.: *Environnement d'exécution pour des services de calcul à la demande sur des grappes mutualisées.* Ph.D. thesis, Université de Grenoble (2012)
9. Ozsu, M.T., Valduriez, P.: Distributed data management: unsolved problems and new issues. Read. Distrib. Comput. Syst. 512–544 (1994)
10. Rosalina, N.H.: An approach of securing data using combined cryptography and steganography. Int. J. Math. Sci. Comput. (IJMSC) **6**(1), 1–9 (2020)

11. Veríssimo, P., Rodrigues, L.: Fundamental security concepts. Distrib. Syst. Syst. Archit. 377–393 (2001)
12. Wang, J., Mu, S.: Security issues and countermeasures in cloud computing. In: Proceedings of 2011 IEEE International Conference on Grey Systems and Intelligent Services, pp. 843–846. IEEE (2011)
13. Yenke, B.O., Mehaut, J.F., Tchuente, M.: Scheduling deadline-constrained checkpointing on virtual clusters. In: 2008 IEEE Asia-Pacific Services Computing Conference, pp. 257–264. IEEE (2008)
14. Bagaza, D.M., Yenke, B.O., Mboliguipa, J.: Triple block data security based on distributed crypto-steganography in the cloud environment. In: 2023 International Journal of Computer Science and Telecommunications, vol. 14, pp. 19–25 (2023)

Author Index

P. Melatagia Yonta et al. (Eds.): CRI 2023, CCIS 2085, p. 191, 2024.
https://doi.org/10.1007/978-3-031-63110-8

Printed in the United States
by Baker & Taylor Publisher Services